Wetland Birds of the Central Plains:

South Dakota, Nebraska and Kansas

Paul Johnsgard

Abstract

This 100,000-word monograph summarizes the distribution, abundance and breeding biology of the 183 species of wetland-adapted birds reliably reported from South Dakota, Nebraska and Kansas through 2011. These include 91 species known to breed or have historically bred in the region, 51 species that migrate through the region but are not yet known to breed or have bred there, and 41 species that are extremely rare, probably extinct, or for which evidence as to their current occurrence is questionable. Brief summaries of the breeding biology of all the regionally nesting species are provided, and information for all species is summarized as to seasonal migrations, habitats, and (in most cases) population status. There is an introductory account of the topography, climate and vegetation of the region insofar as these environmental factors influence wetland birds, six regional maps, and more than 500 references.

Wetland Birds

of the

Central Plains

South Dakota, Nebraska and Kansas

Paul A. Johnsgard

School of Biological Sciences
University of Nebraska – Lincoln

Zea E-Books
Lincoln, Nebraska

ISBN 978-1-60962-018-9 paperback
ISBN 978-1-60962-019-6 e-book

Set in Georgia types.
Design and composition by Paul Royster.

Zea E-Books are published by the
University of Nebraska–Lincoln Libraries.

Electronic (pdf) edition available online at http://digitalcommons.unl.edu/zeabook/
Print edition can be ordered from http://www.lulu.com/spotlight/unllib

Table of Contents

List of Illustrations

Preface and Acknowledgements

The wetlands of North America's Great Plains region represent one of the most productive and biologically diverse habitats of all our continent's natural gifts. They support more than a fifth of all the native bird species of that region, and represent an even greater proportion of the region's overall avian biomass. Most of these wetlands have now been drained and converted to grainfields, subdivisions, or have otherwise been subjected to environmental changes that have degraded or destroyed their wildlife values. Yet, some remnant wetlands still miraculously remain in state or national refuges, state wildlife management areas, and private nature preserves. In such places, essentially all the historic bird life of the original Great Plains wetlands can still be found.

Those bird species associated with the wetlands of three central Great Plains states, from South Dakota through Kansas, are the subject of this study. I chose these three states because collectively they include nearly all the wetland-dependent birds of the entire Great Plains region, from the ducks and grebes largely associated with the glaciated wetlands northeastern plains to the rails, gallinules and herons typical of the warmer wetlands of the more southern states. Although my primary emphasis is on the wetland birds of Nebraska, as I am most familiar with my adopted state and its birds, the additions of South Dakota and Kansas bring additional species and habitats into the overall ecological picture. A companion description and survey of the wetlands of Nebraska is being separately published (Johnsgard, 2012).

About three decades ago I published a survey of the breeding birds of all the Great Plains states, from North Dakota to northwestern Texas, which was later revised and placed online (Johnsgard, 1979; 2009). Much of the present book is based on that source and my other recent regional writings (e.g., Johnsgard, 2007), but most of the earlier information has been updated and expanded. It is hoped that this contribution will provide a useful assessment of the values of the region's wetlands, and help to recognize more generally the Great Plains' wetlands and their related birds as an important segment of the region's environmental riches.

The material in this monograph is based on the work of many people, most importantly those in Nebraska, South Dakota and Kansas who have documented and reported their bird observations over the past century and more, to which the 500-plus citations amply testify I also thank the University of Nebraska's School of Biological Sciences for continuing to provide me with space and resources that have allowed me space and facilities to conduct research long past my official retirement. I especially wish to thank Paul Royster of the University of Nebraska's Digital Commons program (the Center for Digital Research in the Humanities), which has allowed me easily to recover and update relevant text from my 1979 book on the breeding birds of the Great Plains, and also has provided a convenient place to deposit the present manuscript.

Paul A. Johnsgard

Lincoln, Nebraska
January, 2012

Introduction

Topography and Climate of the Great Plains

Although there are minor variations, the overall topography of the Great Plains region is an inclined plain, which slopes downward from the west to the east at an average gradient of about ten feet per mile (Fig. 1). Over almost the entire three-state region under consideration, the overall drainage pattern is to the southeast, into the Missouri and Mississippi river systems. A minor exception occurs in northeastern South Dakota, where glacial moraines in the vicinity of Lake Traverse have produced a low divide that has shifted the drainage patterns from south to north, directing water initially into the Red River in southern North Dakota, and thence into Manitoba and Hudson Bay.

The pattern of rainfall throughout the Great Plains states is relatively simple (Fig. 2). In general, precipitation increases from northwest to southeast, at the approximate rate of about one inch per 40 miles at the northern edge of the region, to about one inch per ten miles at the southern end. This increasing precipitation toward the south is largely counterbalanced by increased rates of evaporation caused by generally warmer temperatures toward the south. About three-fourths of the rainfall in the Great Plains occurs during the growing season, but meltwaters from winter snows are often important in maintaining seasonal wetlands for wetland birds during the early breeding season.

Evaporation rates increase correspondingly as one proceeds south. The highest rates of annual evaporation occur in western Texas, a region characterized by evaporation rates more than four times greater than precipitation rates. Much lower evaporation rates are typical of the cooler, more northerly states, and parts of eastern South Dakota can thus support a lush tall-grass prairie vegetation with less than 30 inches of precipitation a year. In contrast, the same amount of precipitation on southwestern Kansas allows only for the survival of shortgrass prairies and cactus.

Wind has a strong accelerating influence on evaporation rates during summer, and produces devastating effects on protoplasm in conjunction

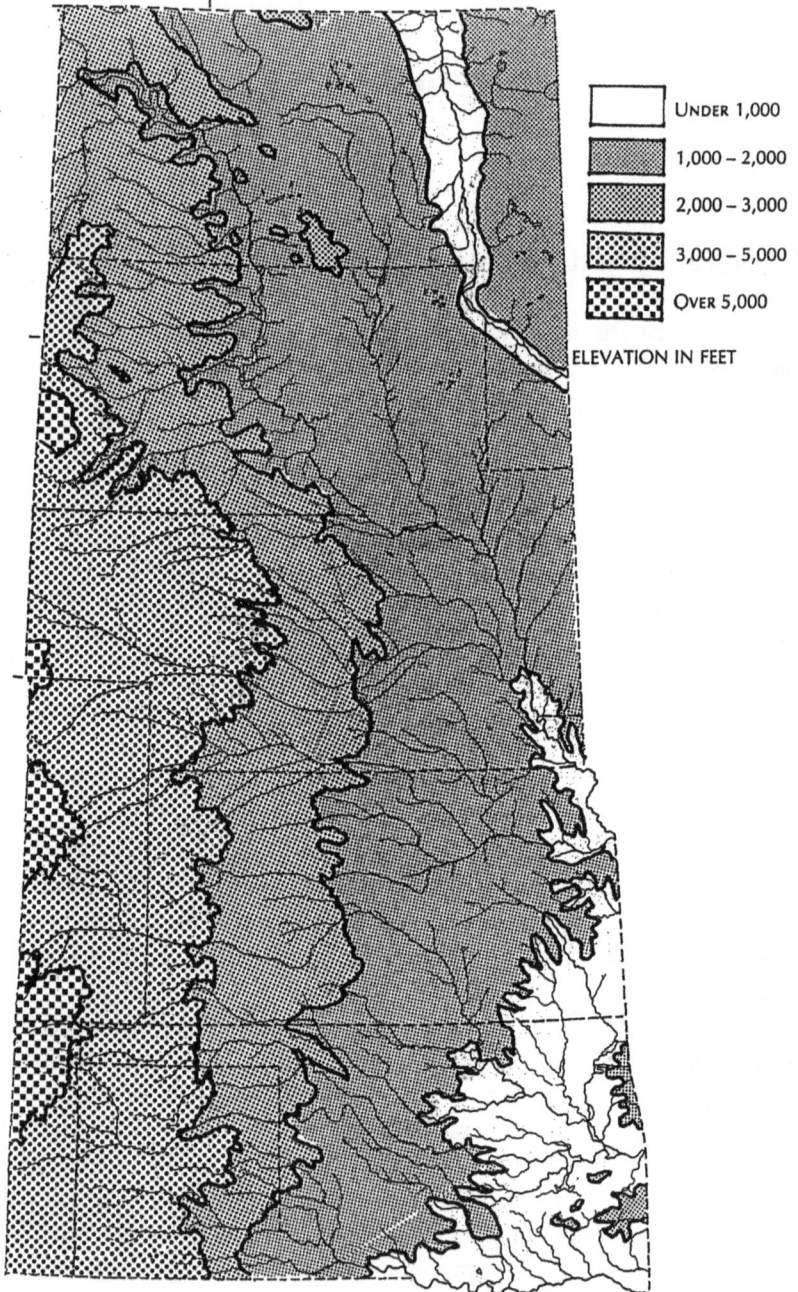

Figure 1. Topography of the Great Plains, adapted from a map in the *Oxford Atlas*.

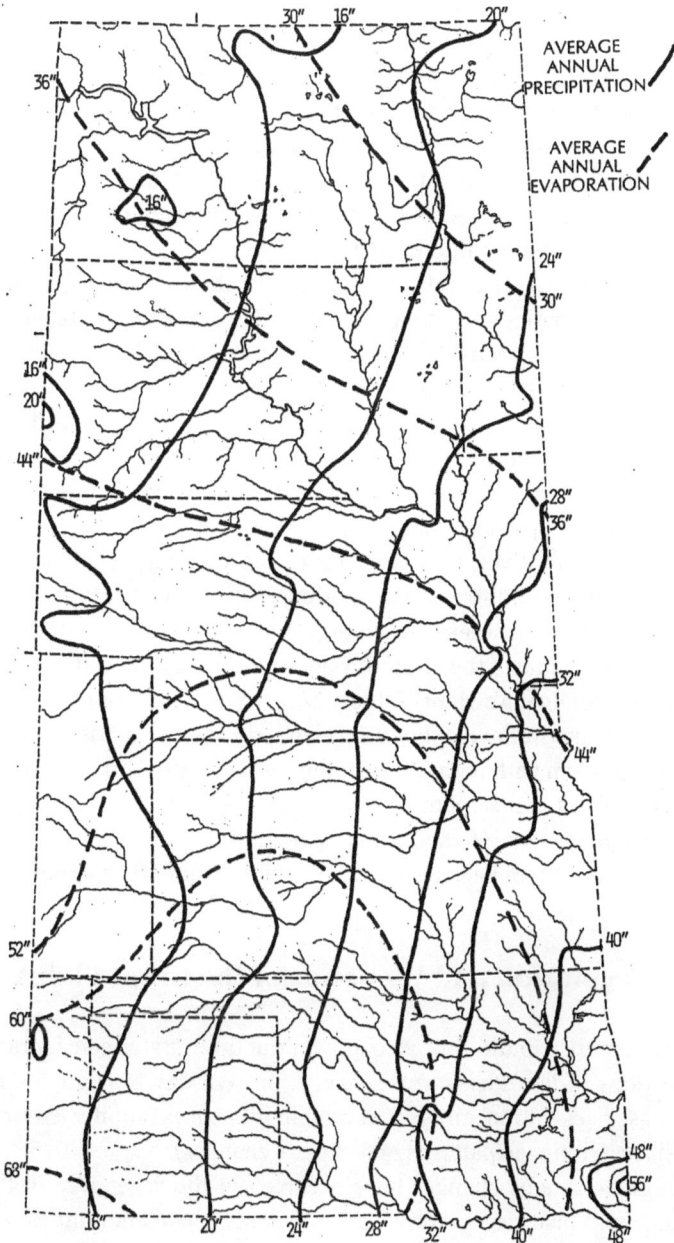

Figure 2. Average annual inches of precipitation (solid line) and annual lake evaporation isopleths (broken line) in the Great Plains, based on U.S. Weather Bureau data.

with sub-freezing temperatures. The Great Plains region includes five of the country's six windiest states, with North Dakota ranking first, followed in sequence by Texas, Kansas and South Dakota. Nebraska is sixth. Southerly winds are usual during late spring and summer, when warm and moist Gulf Coast air drifts north and meets cooler air masses, generating heavy rains, thunderstorms, and occasional tornadoes. In contrast, northerly wind flows are typical during winter, as cold fronts sweep southward out of Canada, often transforming autumn into winter overnight.

Winters in the Great Plains are famous for their severity, bitterly cold winds and long periods of standing snow. They are also notable for the sometimes-sudden blizzards that may materialize with little warning, and may produce periods of little or no visibility for 24 hours or more, together with bone-chilling winds. Freezing rain can be as dangerous to birds as blizzards. During the early 1990's one late-winter storm in Nebraska marked by high winds and freezing rain killed hundreds if not thousands of migrant sandhill cranes. Apparently panicked by high winds, freezing rain and a partly frozen Platte River, they left their roosts and, with ice blinding their eyes, flew headlong into trees and buildings. Crippled and dead cranes littered the landscape for weeks, providing a sudden food source for predators and scavengers. Some of the cripples were captured and moved to the Safari Park division of Omaha's Henry Doorly Zoo, where they formed the nucleus of a sandhill crane wetland exhibit. Likewise, during the 1960s a late winter ice and snow storm with high winds in the central Platte Valley killed or disabled hundreds of cranes. Similarly, a March tornado near York, Nebraska in 1990 swept up tens of thousands of migrating snow geese and Ross's geese, then threw them back to the ground, killing more than a thousand.

Native Vegetation and Wetlands of the Great Plains

The current distributions of wetlands in the three states considered here are reflections of late Pleistocene geology, present-day topography, native climate-based vegetation, and recent human influences (aquifer extractions, modified local drainage patterns and climate changes).

Although enormous changes have occurred in the vegetation of the region, numerous historical records and sufficient relict communities still exist to provide a reasonable basis for understanding the presettlement distribution of vegetation types through the region, which have major effects on the distribution of wetlands. Largely on the basis of the vegetation map assembled by Küchler (1964), it is possible to estimate the relative abundance of major plant communities that once covered the land surface of the region. On that basis, it seems likely that the entire region mapped in figures 1–3 was once at least 80 percent covered by grasslands of varying statures and compositions. Other than a few wetlands associated with wooded riparian wetlands or impounded rivers, wetlands occurring in the region are all associated with non-forested communities, especially grasslands. Except for the wetlands of the Nebraska Sandhills, the region's wetlands are not connected with and dependent upon underground aquifers, but rather are independent and often transitory surface phenomena. They are mostly the result of precipitation-dependent filling of irregular glacial-based topography (the glaciated potholes of eastern South Dakota), or of wind-caused shallow excavations of loess soils (the playa wetlands of Nebraska and Kansas).

Estimates of historic wetland acreages within the three states discussed here averaged about 4.4 percent of the region's total area---Kansas 1.6 percent, South Dakota 5.6 percent, and Nebraska 5.9 percent (Dahl, 1990). In spite of the small area of wetlands now present, wetland birds make up a disproportionately high percent of the region's birds. For example, of Nebraska's approximately 450 bird species, at least 176 (39 percent) are associated with wetlands, so that the state's wetlands are about seven times more species-diverse than what would be expected on the basis of the percentage of Nebraska's land area occupied by them. In Kansas, with at least 425 species and only 1.6 percent of the land represented by wetlands, and at least 162 wetland birds (38 percent), the difference between expected and actual species-diversity is about 24-fold. These differences would probably be even greater if relative biomass of the birds could be calculated; wetlands birds tend to be larger and more numerous than dryland birds of the same region.

It has been estimated that between 1780 and 1980 South Dakota's wetlands were reduced from 2.735 million acres to 1.78 million acres (a 35 percent loss), Nebraska's from 2.91 million acres to 1.9 million (also a 35 percent loss), and Kansas' from 841 thousand acres to 435 thousand (a 48 percent loss) (Dahl. 1990). A statistically significant part of the region's remaining surface water acreage is the result of recent river impoundments, especially the four mainstem dams on the Missouri River, which collectively impound nearly 600 square miles.

The native grassland-dominated communities in the region consist of several associations, ranging from tall-grass prairies to short-grass plains or steppe vegetation (Fig. 3). Of these, the tallgrass prairies at the glaciated

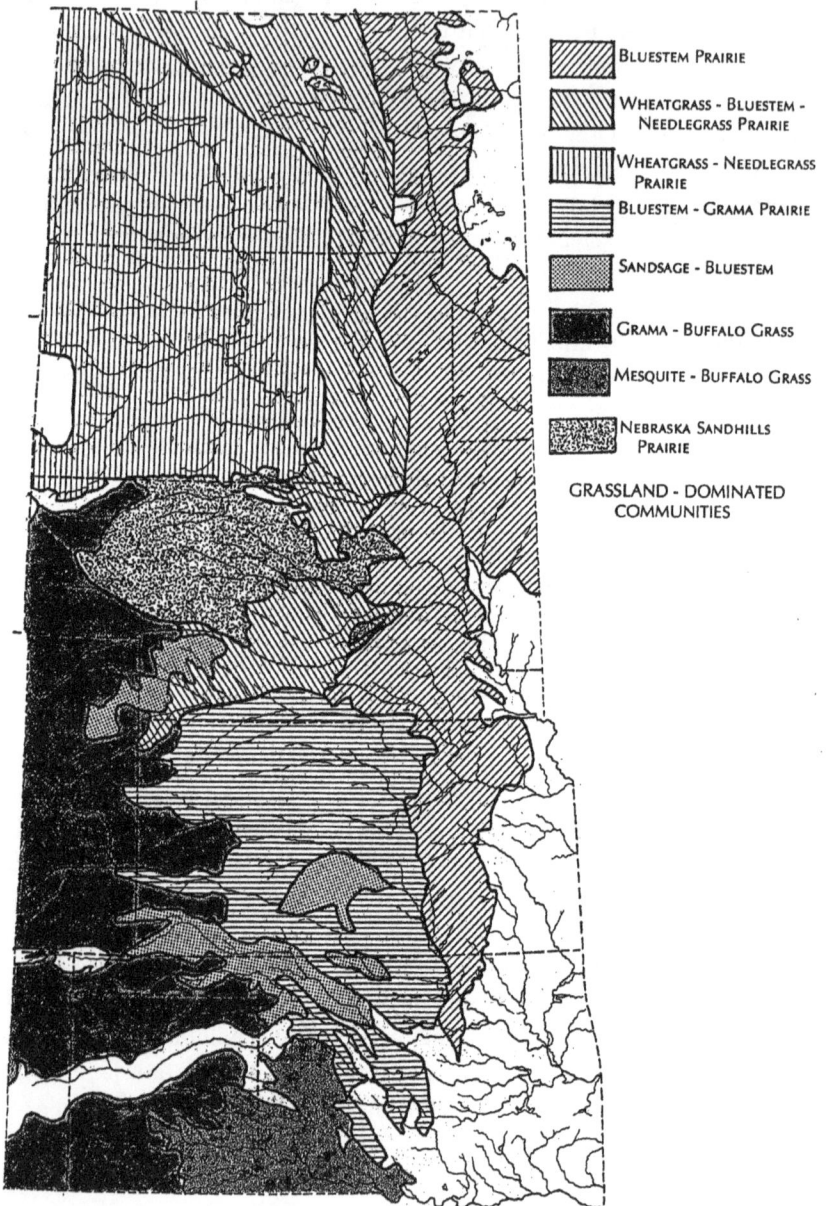

Figure 3. Distribution of non-forested natural plant communities in the Great plains states, showing grassland-dominated vegetation types. Based on Küchler (1964).

eastern edge of the region are the most fragmented and rarest of American prairies, as well as being the most species-rich. Dominated by big bluestem and other tall perennial grasses, these prairies historically covered the eastern Dakotas, western Minnesota and Iowa, plus portions of eastern Nebraska and Kansas, terminating in northern Oklahoma. This entire region was glaciated as recently as 10,000–20,000 years ago. and, except for the table-top flatlands of the Red River valley that were the bottom of glacial Lake Aggassiz, is still rich in prairie marshes and other wetlands (Berry and Buechner, 1993; Hubbard, 1989; Kanrud, Krapu and Swanson, 1989).

To the west of the now-nearly vanished tallgrass prairies in the Dakotas lies the eastern mixed-grass prairie (Stewart 1975) or the wheatgrass-bluestem-needle-grass prairie (Küchler, 1964). The dominant plants are shorter than those of tallgrass bluestem prairie, but a large number of flowering forbs are also characteristic. At least to the Missouri River valley, this region too was glaciated, and is an important part of the recently glaciated "pothole country," or "duck factory" that extends south from the Prairie Provinces of Manitoba and Saskatchewan across eastern North Dakota to the southeastern corner of South Dakota.

The wetlands of South Dakota are almost entirely confined to the state's glaciated regions east of the Missouri River (Johnson *et al.*, 1997). Wetlands cover about 9.8 percent of the land in eastern South Dakota, 60 percent of which are less than half an acre in area, and less than five percent are larger than five acres. They were formed from glacial till deposited at the end of the last (Wisconsinan) glacial period about 10,000 years ago. As of the mid-1990s, these wetlands covered an estimated 2,137,900 acres. More than half of them (over 520,000) are temporary, sometimes lasting only a few weeks. There are also about 334,00 seasonal wetlands, covering 553,500 acres and average less than two acres in area, which typically become dry by mid-summer. The county with the largest number of seasonal wetlands is Edmunds County, with 22,225, while Union County has only 998.

There are over 11,800 semipermanent wetlands in Marshall, Day, Roberts, Grant and Deuel counties, and represent about half of eastern South Dakota's total of 23,997, covering 377,000 acres. These may during wet years may remain wet throughout the year, and in wet years may expand and function like lakes. Semipermanent wetlands comprise 2.5 percent of all the wetlands in eastern South Dakota, but are especially valuable for species that need access to water throughout the summer breeding season.

There are also over 600 permanent wetlands covering some 194,000 acres in eastern South Dakota, which are located in a broad north-south belt of glacial till extending from Day County south to Lake County. Lake Poinsett in Brookings and Hamlin counties is the state's large natural wetland, of 10,000 acres. Six counties in northeastern South Dakota--- Beadle, Day, Hamlin, Kingsbury, Roberts and Sanborn—have from 12–16 percent of their areas occupied by wetlands.

There are also over 77,000 created wetlands in eastern South Dakota, including 56,827 "dugouts" mostly formed by excavating natural depressions such as natural wetlands (62 percent), in stream channels, or elsewhere. There are 11,838 stock dams and 75 reservoirs in eastern South Dakota that respectively cover 106,740 and 256,000 acres. The smaller stock dams provide relatively more benefits to wildlife per acre of water than do the large reservoirs.

In eastern South Dakota about 75 percent of the remaining wetlands have no legal protection from damage or destruction (Johnson *et al.*, 1997; Johnson and Higgins, 1997). About 465,000 wetlands, or about half the total, are especially threatened by proposals to eliminate protection for frequently farmed sites and wetlands less than an acre in area (Johnson *et al.*, 1997).

In several areas, extensive regions of sandy soil or sand dunes have greatly affected the distributions and types of native vegetation. The largest of these is the Nebraska Sandhills, where the vegetation consists mostly of widely spaced bunchgrasses, with the intervening areas very sparsely vegetated. In spite of the surface aridity, the Sandhills lie atop the Ogallala aquifer. This zone of saturated sand and gravel is the largest aquifer in the Western Hemisphere, and is more than 600 thick over much of the region (Bleed and Flowerday, 1989). In low, interdune depressions, wet meadows and marshes are formed where the top of the Ogallala aquifer reaches the sand surface, produces thousands of relatively small and shallow wetlands of great value to wetland birds (Novacek, 1989; Johnsgard, 1995; 2001a). Most of these are less than an acre or so in area, and all are shallow. The largest Sandhills lake is Cottonwood Lake in Cherry County, which is only 20 feet deep, and the second largest, Crescent Lake in Garden County, is only 16 feet deep.

The Sandhills lakes also very greatly in water chemistry. They range from nearly neutral in pH, in central and eastern regions, to highly alkaline in the western Sandhills. An estimated total of 394,000 acres of wetlands having low alkalinity are associated with the main Sandhills region, while the hyperalkaline wetlands in the western Sandhills comprise another 10,700 acres. A small region of sandhill wetlands in the Loup–Platte Valley add another 8,000 acres (LaGrange, 2005). The hydrology of the Sandhills was thoroughly described in Bleed and Flowerday's comprehensive monograph (1989), and the ecology of the Sandhills wetlands and their biotas have been discussed by Johnsgard (2012).

The wheat-grass-needlegrass, or western mixed-grass, prairies historically occupied nearly all of North Dakota from the Missouri Valley westward and extended over more than half of South Dakota. The native vegetation is predominantly short-grass species and scattered midgrasses, plus a moderate number of broad-leaved forbs. This unglaciated and relatively region is largely devoid of statistically significant wetlands except for those associated with rivers.

The short-grass prairie, or grama-buffalo grass association, occurs on localized slopes and dry exposures in the western Dakotas and over extensive portions of the region from western Nebraska southward to the arid Staked Plains of northwestern Texas. This "high plains" biota is adapted to withstand considerable aridity, and its array of both plants and animals is somewhat restricted. Surface wetlands in this region are both rare and temporary, and typically are the result of periodic heavy rains. Much of this region depends on snowfall and spring rains to fill temporary wetlands in windblown depressions, producing shallow playa wetlands (Steiert, 1985; Smith, 2003).

These temporary playa wetlands occur from western Texas north to northwestern Nebraska, a region that depends largely on winter and spring precipitation for annual recharge.. They are often rich in invertebrate life and thus are highly valuable as foraging sites and migratory stopover points for migratory wetland birds (McCrae, 1972; Cariveau, Johnson and Sparks. 2007; Cariveau and Pavlacy, 2009). LaGrange (2005) estimated that playa wetlands in Nebraska's Rainwater Basin totaled some 34,000 acres as of the early 2000's, those in southwestern Nebraska 21,680 acres, and those in the central tablelands 7,317 acres. Probably 90 percent of the historic Rainwater Basin wetlands had been lost to drainage and agricultural conversion by the early 1980's (LaGrange, 2005).

The general regional locations of major wetland types in the three-state region is shown in Fig. 4. For convenience in locating sites mentioned in the text, maps showing the names of counties in each of the three states are also provided (Fig. 5).

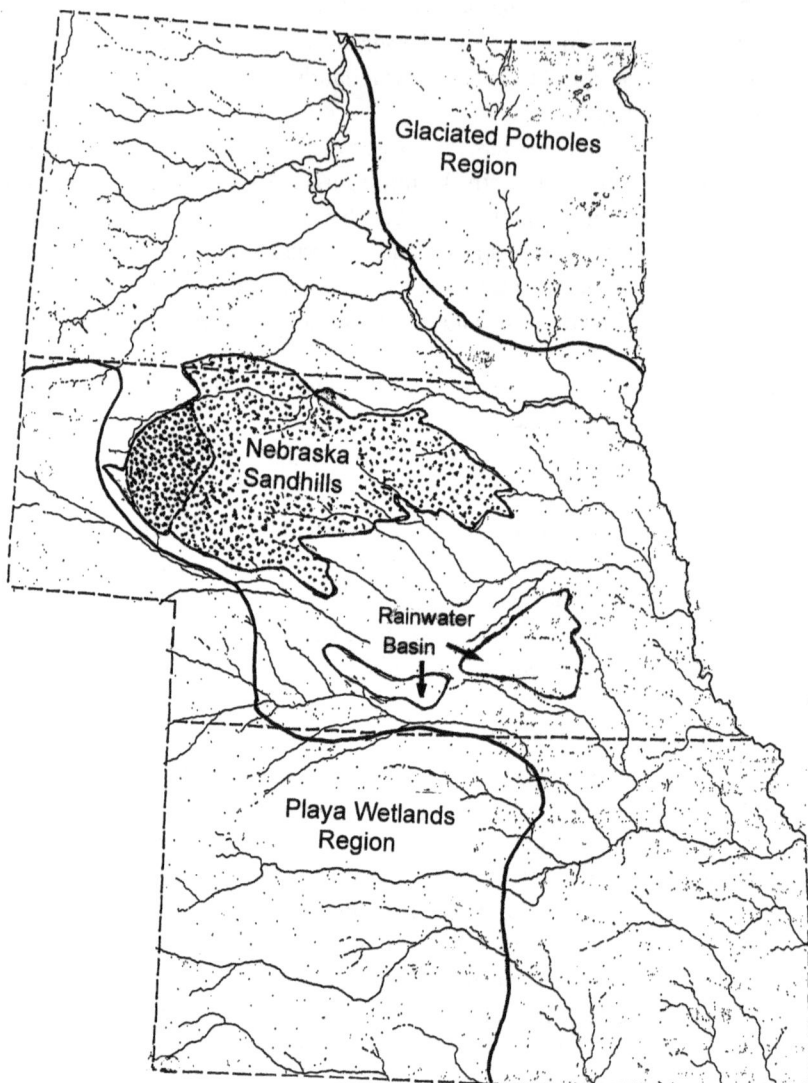

Figure 4. Distribution of major wetland regions in the central Great Plains. Heavier stippling in Sandhills indicates alkaline wetlands region. Map by author, based on various sources.

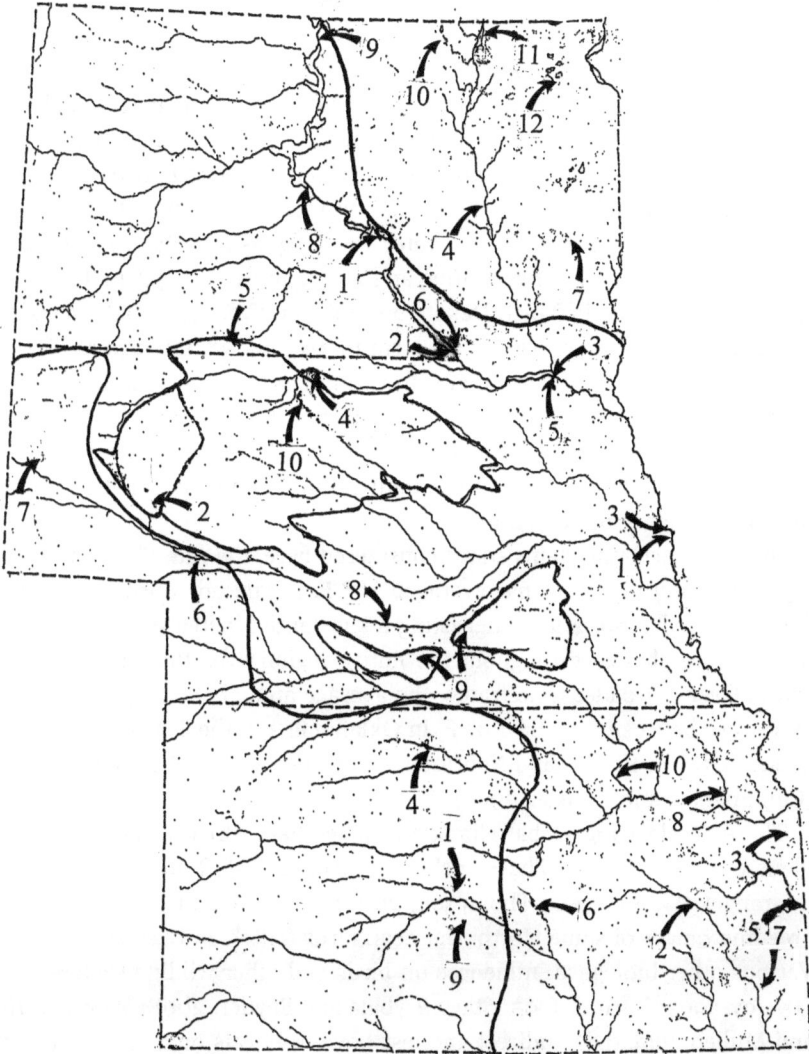

Figure 5. Locations of important regional habitats for wetland birds in the central Great Plains states. See pages 22–34 for site identifications.

Recent Climate Change in the Great Plains

For much of the 20th century, and especially during the past few decades, the Great Plains have experienced a warming trend that is part of a global phenomenon. In the Great Plains this warming trend has been most apparent in recent milder winter temperatures, which are more noticeable in the northern states such as the Dakotas than in more southern latitudes. Thus the century-long average rate of winter warming, as reflected in statewide January temperatures between 1895 and 2008, averaged 0.44° F. degree of increase per decade in North Dakota but only 0.04° F. per decade in Oklahoma, a ten-fold difference in long-term mid-winter warming rates (Table 1). This increasing influence of global warming is most evident at extreme northern and southern latitudes, where the melting of permafrost and disappearance of ancient glaciers, are obvious. At lower latitudes increased droughts and record high summer temperatures are increasingly uncomfortable aspects of modern life.

In the three Great Plains states considered here, the rate of mean January temperature increase has averaged 1.7° F. per decade during the four-decade period 1969–2008, suggesting a substantial increase in the rate of temperature change throughout the entire region over the past four decades. Over a broader time-frame, the average annual 1895–1994 temperature rose between 0.5 and 1.0° F. in Oklahoma and the Texas panhandle, about 1.0° F. in Kansas and Nebraska, and from 1.5–3.0° F. in the Dakotas (Cunningham and Kroeger, 1996).

These accelerating global climatic changes have already had many evident effects on birds. These include a poleward shift in avian wintering ranges (La Sorte and Thompson, 2007), northward movements in the breeding ranges of some North American birds (Hitch and Leeberg, 2007). Various other biological influences on birds and other wildlife (Peters and Lovejoy, 1992; Burton, 1995; Stavy, Dybala and Snyder, 2008; Wormworth and Mullen, undated). Indirect and less obvious effects on a species might result from climate-based influences on its parasites, diseases, competitors and predators. Still other more dramatic effects of global warming, such as changes in the frequencies of droughts, floods, rainstorms, hurricanes and other climatic disasters, may have massive if short-term consequences on local or regional populations. These additional influences might, for example, include measurable changes in a species' breeding phenology or fecundity, in the composition and structure of its breeding and wintering habitats, or its migration timing, routes and staging areas.

Table 1. Mean January Temperatures (Fahrenheit), 1969–2008*

	1969–1978	1979–1988	1989–1998	1999–2008	1895-2008 average*
S. Dakota	11.8°	16.9°	17.4°	20.7°	15.56°
Nebraska	19.5°	22.3°	24.55°	25.7°	22.53°
Kansas	26.4°	26.3°	30.9°	32.0°	28.92°
3-state ave.	19.2°	21.8°	24.6°	26.1°	20.03°

* U.S. Weather Bureau data

In North America the average spring arrival time of many short-distance migrants breeding in the Northeast occurred an average of nearly two weeks earlier during the second half of the 20th century than the first half (Butler, 2004). Over a 63-year period of the 20th century (1939–2001), at least 27 migratory species exhibited altered spring arrival dates at Delta Marsh, Manitoba, with 15 of the species arriving significantly earlier as the century progressed (Murphy-Klassen et al., 2005). Mills (2005) found that both spring and fall migration patterns of passerines at Long Point, Ontario, were affected by global warming during the period 1975–2000. Fall migrations were especially affected, with 13 of 14 species studied exhibiting delayed fall migrations.

Snow geese have shifted from wintering on the Gulf Coast of Texas to refuges as far north as Kansas, while Canada geese are also now commonly wintering in northern Kansas and Nebraska, and locally into the Dakotas. Double-crested cormorants, which at the time of Root's 1963–1972 analysis barely appeared on Oklahoma's Christmas Counts, have increased a thousand-fold in average numbers seen there. In the fall and winter of 2011-2012, thousands of sandhill cranes remained on Nebraska's central Platte valley well into January, as did several hundred thousand snow geese at Squaw Creek National Wildlife Refuge in northwestern Missouri.

One result of such changes is that published breeding and wintering ranges of many species that are more than a decade or two old are now increasingly inaccurate. Likewise, average spring and fall migration arrival and departure dates that are based on data several decades old are similarly suspect. Those determined for Nebraska for a 50-year period from the 1930's to 1980 by Johnsgard (1980a, 1980b) and used in this report, are now clearly outdated. They need to be adjusted anywhere from 1–2 weeks (in spring) or 2–4 weeks (in fall) to conform to current migratory phenology patterns. They probably more closely reflect the current phenologic migration pattern in South Dakota.

In Nebraska, and Kansas many water-dependent species are now wintering commonly on ice-free rivers and impoundments. The largest recent Christmas Count numerical increases in the Lincoln count circle have occurred among Canada geese and mallards, with Canada geese now regularly being the most numerous species reported. During the decade 1966-67–1976-77, Canada geese were reported an average of 0.8 birds per party hour in state-wide Nebraska counts, but by the decade 1996-97–2006-07 the state-wide average had increased to 81.8 birds per party hour. Over that period, maximum region-wide average numbers for Canada goose have shifted from northwestern Texas to Kansas, and for mallards maximum regional averages have similarly moved north from Oklahoma to Kansas. Many other wetland and terrestrial species have similarly shifted their early-winter populations northward during this four-decade period (Johnsgard and Shane, 2009).

Additionally, some wetland species that rarely or never appeared on early Christmas counts in northern areas have been increasingly seen in recent counts. For example, since 1998 three species of ducks and three sandpipers, as well as the western grebe and marsh wren, have all appeared on Lincoln's Christmas Bird Counts for the first time, and similar trends have occurred at Scotts Bluff, in western Nebraska (Johnsgard, 1998, 2006).

Important Regional Habitats for Wetland Birds

Note: The sites described below are listed alphabetically and numbered for each state, and their approximate geographic locations are shown numerically in Figure 5. Figure 6 shows the distribution and names of counties in the three states.

South Dakota

1. Big Bend Dam and Lake Sharpe. This Corps of Engineers flood-control dam in Buffalo and Lyman counties impounds Lake Sharpe, a flood-control reservoir of 57,000 acres. Like other large Missouri River reservoirs, Lake Sharpe often attracts large numbers of migrating waterfowl and gulls. No bird list is yet available. Address of Big Ben Dam: U.S. Army Corps of Engineers Big Bend Project, 33573 N. Shore RD., Chamberlain, SD 57325 (ph. 605-245-2255).

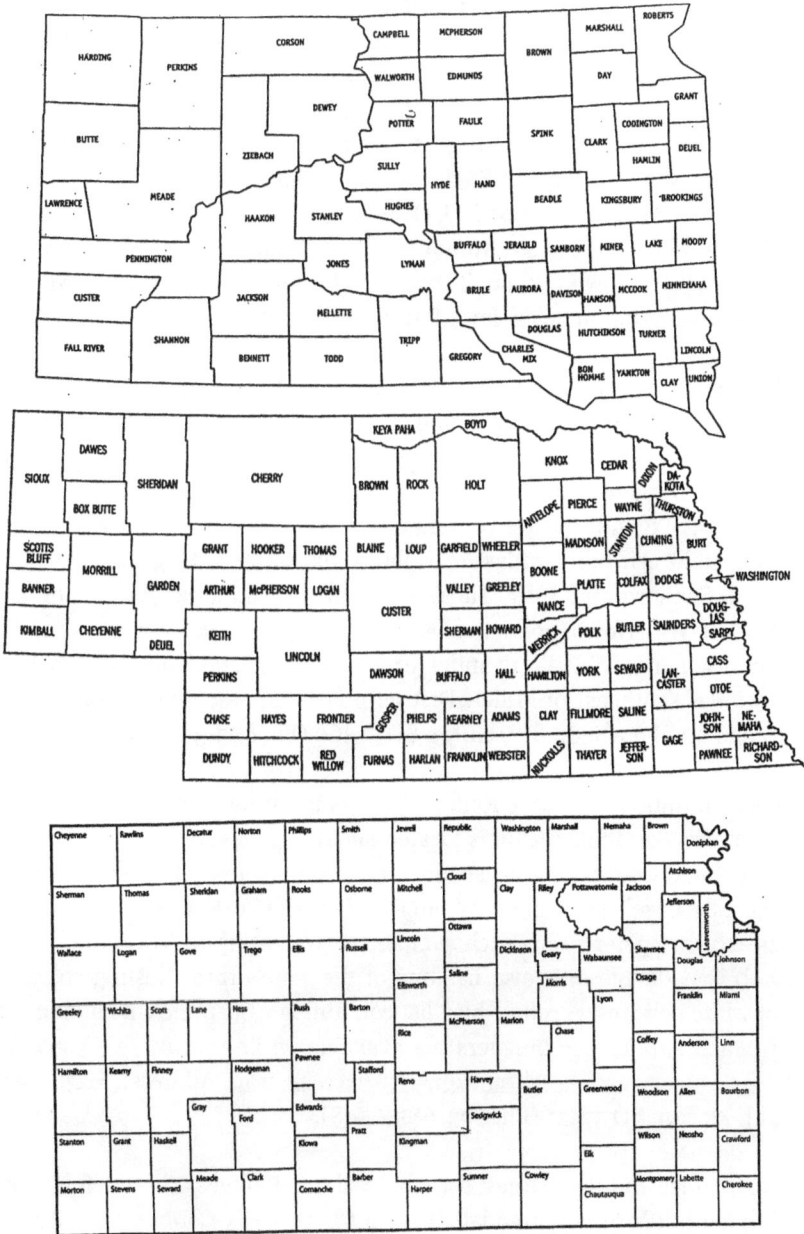

Figure 6. County names and locations in South Dakota (top), Nebraska (middle), and Kansas (bottom).

2. Ft. Randall Dam and Lake Francis Case. This Corps of Engineers flood-control dam in Charles Mix and Gregory counties impounds Lake Francis Case, a flood-control reservoir of 102,000 acres. Like other large Great Plains reservoirs, this lake often attracts large numbers of migrating waterfowl and gulls. No bird lists are yet available. Address of Ft. Randall Dam: U.S. Army Corps of Engineers Fort Randall Project, U.S. Hwy 281 & 18 399 Powerhouse Road. P.O. Box 199, Pickstown, SD 57367-0199 (Ph. 605-487-7845). Karl Mundt National Wildlife Refuge (780 acres) is located just below Ft. Randall Dam; see the Lake Andes National Wildlife Refuge description below for information on both.

3. Gavin's Point Dam & Lewis & Clark Lake State Recreation Area. See description and contact information in Nebraska section below.

4. Huron Wetland Management District. This wetland district manages more than 87,5000 acres of wetlands in 62 waterfowl production areas, over eight central counties---Beadle, Buffalo, Hand, Hughes, Hyde, Jeraud, Sanborn and Sully. Three excellent wetland areas for birding are LeClair Waterfowl Production Area (W.P.A.), 13 miles northwest of Iroquois, Bauer W.P.A., 13 miles east of Huron, and Campbell W.P.A., 15 miles southeast of Miller. District office address: Room 308 Federal Bldg., 200 Fourth St. SW, Huron SD 57360 (Ph. 805-353-5894).

5. Lacreek National Wildlife Refuge. This refuge at the northern edge of the Nebraska Sandhills region is located about 15 miles southeast of Martin, in Bennett County. It consists of extensive marshes and shallow lakes in the Lake Creek valley of the White River's South Fork. The refuge hatches 15–20 trumpeter swans each summer, as well as up to 6,000 ducks and 800 Canada geese. It also has one of the state's three nesting colonies of white pelicans. A refuge bird list containing 273 species, including 117 wetland species (46 breeders), is available on-line: http://www.npwrc. usgs.gov/resource/birds/chekbird/r6/lacreek.htm. Address: HC5, Box 114, Martin, SD 57551 (Ph. 605/685-6508).

6. Lake Andes and Karl Mundt National Wildlife Refuges. These refuges are located north of Fort Randall Dam in Charles Mix County. Lake Andes N.W.R. consists of 5,450 acres around Lake Andes, a shallow Pleistocene glacial lake and marsh, including about 4,700 acres of open water and marsh. Karl Mundt National Wildlife Refuge is a small refuge (780 acres)

established for bald eagles, located directly below Ft. Randall Dam. The refuge is not open to the public, but an eagle observation platform on U.S. Corps of Engineers property at Fort Randall Dam is provided. The state-controlled Lake Andes Wetland Management District encompasses 34,682 acres of wetlands and grassland easements over a 13-county area of southeastern South Dakota, including Aurora, Bon Homme, Brule, Charles Mix, Clay, Davison, Douglas, Hanson, Hutchinson, Lincoln, Turner, Union and Yankton counties. A bird list for the Lake Andes complex (including both refuges and the wetland management district), has 213 species, with 97 wetland species (37 breeders), and is available on-line: http://www.npwrc.usgs.gov/resource/birds/chekbird/r6/lakeande. htm. Address of both refuge and district office: 38672 291 St., Lake Andes, SD 57456 (Ph. 605-487-7603).

7. Madison Wetland Management District. This wetland district in the heart of the glaciated pothole region manages more than 52,000 acres of wetlands, ranging in size from 40–400 acres, and including over 36,000 acres preserved in waterfowl production areas. It includes eight east-central counties: Brookings, Deuel, Hamlin, Kingsbury, Lake, McCook, Minnehaha, Moody, Sanborn and Sully. A bird list containing 297 species, including 96 wetland species (42 breeders) that have been observed in the district, is available on-line: http://www.npwrc.usgs.gov/resource/ birds/chekbird/r6/madison.htm. Address: Madison Wetland District Office, Box 48, Madison, SD 57042 (Ph. 605-256-2974).

8. Oahe Dam & Lake Oahe. This Corps of Engineers flood-control dam and largest of the regional Missouri River reservoirs impounds 206,000 acres of water. Like the other Missouri River reservoirs, Lake Oahe often attracts large numbers of migrating waterfowl and gulls. Address of Oahe Dam: U.S. Army Corps of Engineers, 28563 Powerhouse Rd., Pierre, SD 579501 (Ph. 605-224-5862.

9. Pocasse National Wildlife Refuge. Area 2,540 acres, just north of Pollock, Brown County, off U.S Hwy. 83, and bordering the east side of the Missouri River. Mostly marshes and open water (1,045 acres of wetland), this refuge is an important stopover area for migrating sandhill and whooping cranes, as well as for waterfowl. No separate bird list for Pocasse N.W.R. is yet available, but the Sand Lake N.W.R. list may be applicable. Address: Administered out of Sand Lake N.W.R. (Ph. 605-885-6320).

10. Samuel H. Ordway Jr. Memorial Prairie Preserve. Located northwest of Aberdeen near Leola, in McPherson County, this 7,800-acre preserve of mixed-press prairie has about 400 prairie potholes that attract migrating and nesting shorebirds and waterfowl. No bird list is yet available. Address: HCR 1, Box 26, Leola SD 57456 (Ph. 605-439-3475).

11. Sand Lake National Wildlife Refuge. This 45,000-acre refuge is located 25 miles northeast of Aberdeen, Brown County, in the James River Valley, and was part of the shoreline of glacial Lake Dakota until about 10,000 years ago. It consists of more than 21,000 acres of marshes, grasslands, shallow impoundments, shelterbelts and fields. Sand Lake N.W.R. has the world's largest nesting colony of Franklin's gulls, and has been identified as a wetland of international importance. It attracts hundreds of thousands of snow geese and other waterfowl during migration, and attracts four breeding grebe species (eared, western, Clark's and pied-billed), as well as nesting canvasbacks, redheads, lesser scaups and ruddy ducks. Notable nesting shorebirds include the marbled godwit, Wilson's phalarope and three terns (common, Forster's and black) The refuge also administers Pocasse National Wildlife Refuge (see above), and is located among more than 150,000 acres of regional state-owned wildlife management areas in the glaciated pothole region of northeastern South Dakota. The Sand Lake Wetland Management District is the largest wetland management district in the country, encompassing 9,000 square miles. It includes ten of South Dakota's north-central counties: Brown, Campbell, Corson, Dewey, Edwards, Faulk, McPherson, Potter, Spink and Walworth. It contains 45,000 acres of land under federal protection, involving 162 waterfowl production areas, and includes an additional 550,000 acres protected by conservation easements. A bird list for Sand Lake N.W.R. totals 263 species, including 106 wetland species (of which 55 are breeders, the largest number for any regional site), and is available on-line: http://www.npwrc.usgs.gov/resource/birds/chekbird/r6/sandlake.htm. Address of Sand Lake Refuge: R.R. 1, 399650 Sand Lake Drive, Columbia, SD 5433 (Ph. 605-885-6320).

12. Waubay National Wildlife Refuge. This 4,600-acre refuge is situated eight miles north of Waubay, Day County, in the glaciated till region of northeastern South Dakota, the heart of South Dakota's glacial pothole country. It contains nearly 5,000 acres of marshlands, lakes, grass-

lands, brush, and oak woodlands. It has all the species of nesting grebes as those mentioned for Sand Lake National Wildlife Refuge, plus horned and, occasionally, red-necked grebes. It also attracts the same nesting species of diving ducks, shorebirds and terns. A refuge bird list containing 244 species, including 103 wetland species (52 breeders), is available on-line: http://www.npwrc.usgs.gov/resource/birds/chekbird/r6/waubay.htm. Pickerel Lake State Recreation Area (55 acres) is just north of the refuge. It is developed for fishing, but also attracts waterfowl during migration. The Waubay Wetland Management District includes over 300 waterfowl production areas totaling some 4,000 acres in six wetland-rich northeastern counties: Clark, Codington, Day, Grant, Marshall and Roberts. Address of refuge and district wetland office: 44401 134A St., Waubay, SD 57273 (Ph. 605-947-4521).

Nebraska

1. Boyer Chute National Wildlife Refuge. Area 3,100 acres. This refuge consists of riparian woods and restored lowland prairie along three miles of the Missouri River, and a restored 2.5-mile restored chute (river channel cut-off). Located three miles east of Fort Calhoun, Washington County. The area and its birds have been described by Farrar (2004) and Johnsgard (2011). No bird list is yet available, but the list for nearby DeSoto N.W.R. is applicable. Address: Administered from DeSoto National Wildlife Refuge (Ph. 402-468-4313). URL: http://midwest.fws.gov/desoto/boyerbro.html,

2. Crescent Lake National Wildlife Refuge. Area 45,818 acres. There are about 20 wetland complexes on this enormous Sandhills refuge; the wetlands total 8,251 acres, and comprise almost 20 percent of the refuge's area. Notable nesting species include cinnamon teal, redhead, canvasback, ruddy duck, black-necked stilt, American avocet and white-faced ibis. The area and its birds have been described by Farrar (2004) and Johnsgard (1995, 2011). The refuge bird list contains 273 species, including 111 wetland species (43 breeders), and is available on-line: http://www.npwrc.usgs.gov/resource/birds/chekbird/r6/crescent.htm. Located 28 miles north of Oshkosh, Garden County. Address: 10630 Rd. 181, Ellsworth, NE 69340 (Ph. 308-762-4893). URL: http://www.lpsnrd.org. http://crescentlake.fws.gov/.

3. DeSoto National Wildlife Refuge. Area 7,823 acres. This refuges includes mature riverine deciduous forest along the Missouri River, a seven-mile-long oxbow lake (DeSoto Lake), and mostly cultivated uplands. Located five miles east of Blair on U.S. Highway 20, partly in Washington County, and partly on the Iowa side of the Missouri River. Breeding shorebirds include piping plover, least tern, and American woodcock. The area and its birds have been described by Farrar (2004) and Johnsgard (2011). The refuge bird list contains 240 species, including 102 wetland species (16 breeders), and is available on-line: http://www.npwrc.usgs.gov/resource/birds/chekbird/r3/desoto.htm Daily admission fee. Address: Rte. 1., Box 114, Missouri Valley, IA (Ph. 712-642-4121). URL: http://midwest.fws.gov/desoto/dsotobro.html.

4. Fort Niobrara National Wildlife Refuge. This refuge along the Niobrara River covers 19,122 acres, including 375 acres of wetlands. Located about four miles east of Valentine, Cherry County, on State Highway 12. This is a good place to observe breeding long-billed curlews and upland sandpipers. The area and its birds have been described by Farrar (2004) and Johnsgard (1995, 2011). The refuge bird list comprises some 230 species, including 81 wetland species (17 breeders), and is available on-line: http://www.npwrc.usgs.gov/resource/birds/chekbird/r6/niobrara.htm. URL: http://fortniobrara.fws.gov/. Address: Hidden Timber Star Route, Valentine, NE 69201 (Ph. 402-376-378).

5. Gavin's Point Dam and Lewis and Clark Lake State Recreation Area. Area of SRA. 33,227 acres, with a 32,000-acre flood-control reservoir on the South Dakota–Nebraska border. Located along the South Dakota boundary, seven miles north of Crofton. No on-line bird list is yet available, but the Lewis & Clark Visitor Center may be able to provide a local bird list. Park permit required for entering State Recreation Area. Address of Gavin's Point Dam: Gavin's Point Project, US Army Corps of Engineers, PO Box 710, Yankton, SD 5707 (Ph. (402-667-7873) Lewis & Clark Visitor Center: Ph. 402-667-2546. URL: http://www.nwo.usace.army.mil/html/Lake_Proj/gavinspoint/recreation.html

6. Lake McConaughy and Lake Ogallala State Recreation Areas (SRAs) Lake McConaughy S.R.A. occupies 41,192 acres, most of which consists of a reservoir about 22 miles long, three miles wide, and with over 100 miles of shoreline when full. Lake Ogallala S.R.A. totals 659 acres, of which

Lake Ogallala comprises 340 acres. This shallow lake is maintained by spillway water from Kingsley Dam, and has large daily and seasonal water fluctuations that greatly affect wetland bird breeding success. Both lakes are managed for irrigation and power by the Central Nebraska Public Power and Irrigation District (308-284-2332, or 284-3542), although Nebraska Game & Parks controls the associated recreation and wildlife management areas. A free-admission bald eagle viewing center below Kingsley Dam and overlooking Lake Ogallala is open from late December to early February; peak eagle numbers of up to several hundred birds (368 being the record total) often occur here during late January or February (Peyton 2012). Kingsley Dam is located nine miles north of Ogallala on State Highway 61. A state park entry permit is required for both sites. Large breeding populations of piping plovers and least terns breed along the lake's sandy shores, and over 100 species of birds breed in the general vicinity. At the western end of Lake McConaughy, Clear Creek Wildlife Management Area provides a large area of wet meadows and other wetland habitats that attract rails, bitterns, and other wetland species. Western and Clark's grebes both breed here, and fall migrant populations of these spectacular grebes sometimes number in the tens of thousands on Lake McConaughy. A recent bird list for Lake McConaughy, Lake Ogallala, and adjoining areas has 363 species, of which more than 150 are wetland species (Brown, Dinsmore & Jorgensen, 2012). This region is notable for the many species of waterfowl (38), gulls (18), and herons and ibises (14) that have been seen here, as well as all three North American jaegers and four loons. Birding opportunities in the region have been described by Farrar (2004), Johnsgard (2011) and Dinsmore (2012). Address of Lake McConaughy S.R.A. & Lake McConaughy Visitor Center, 1475 NE Hwy 61 N, Ogallala, NE (Ph. 308-284-8800), URL: http://www.lakemcconaughy.com/ngp.html.

7. North Platte National Wildlife Refuge. Area 5,047 acres. Part of the Crescent Lake/North Platte N.W.R complex, and including Lake Alice (1,377 acres when full, but recently dry), Lake Minatare (430 acres) and Winters Creek (700 acres). The best wetland bird habitat is at Winters Creek. Located four miles north and eight miles east of Scottsbluff, Scotts Bluff County. The area and its birds have been described by Farrar (2004) and Johnsgard (2011). The refuge bird list totals 228 species, including 85 wetland species (13 breeders), and is available on-line: http://www.npwrc.usgs.gov/resource/birds/chekbird/r6/noplatte.htm.

Address: 10630 Road 181, Ellsworth, NE 69340 (Ph. 308-762-4893). URL: http://crescentlake.fws.gov/northplatte/.

8. Platte Valley. The central section of the Platte Valley, especially the section from North Platte east to Grand Island, attracts up to ten million waterfowl and a half-million sandhill cranes each spring, plus such rare species as whooping cranes, least terns and piping plovers. It is recognized as critical habitat for the federally endangered whooping crane, and is of international importance as a spring staging area for migrating sandhill cranes. In March the region's waste corn typically supports up to about 90 percent of the world's lesser sandhill cranes, as well as up to about nine million other waterfowl. Information on cranes and crane viewing is available locally from the Lillian Annette Rowe Sanctuary & Iain Nicolson Audubon Center (Ph. 308-468-5282) near Gibbon, and the Nebraska Nature and Visitor Center (Ph. 308-382-1820), at the Interstate 80 exchange 305 near Alda. Another local information source for Platte Valley birds is the Kearney office of the Nebraska Game & Parks Commission, 1617 1st Ave., Kearney, NE 68847 (Ph. 308-865-5310). The valley's ecology and birds, especially the waterfowl and cranes, have been described by Krapu (1996), Currier, Lingle & VanDerwalker (1985) and Johnsgard (2008). Lingle (1994) provided a monthly bird list of 300 regional species, including 125 wetland species. A similar list specifically applicable to the central Platte Valley, along with a summary of the region's natural history, is also available on-line (Johnsgard, 2008): http://digitalcommons.unl.edu/biosciornithology/40.

9. Rainwater Basin. This loess-mantled region south of the Platte River contains hundreds of temporary to seasonal playa wetlands, and is geographically divided into eastern and western components. The Rainwater Basin Joint Venture coordinates the Rainwater Basin's wetland management, which involve the approximately 50 federally owned waterfowl production areas, and about 30 state-owned wildlife management areas extending from Phelps County east to Butler and Saline counties. The Rainwater Basin Joint Venture's address is 2550 N. Diers Ave., Suite L, Grand Island, NE 68803 (Ph. 308/382-8112). The Rainwater Basin's importance to Great Plains migrating shorebirds during April and early May is probably second only in the Great Plains to Cheyenne Bottoms in Kansas, and during wet springs it also often holds millions of migrating geese (mostly snow geese, greater white-fronted geese and Canada geese) during March. The area and its birds have been described by Farrar (2004)

and Johnsgard (2011). The address for the Rainwater Basin Wetland Management District is: U.S. Fish and Wildlife Service, 2610 Ave. Q, P. O. Box 1686, Kearney, NE 68847 (Ph. 308-236-5015). Nebraska's playa wetlands are included within the multi-state Playa Lakes Joint Venture program, which extends from northwestern Nebraska south to western Texas. The Playa Lakes Joint Venture's address is 103 East Simpson St., LaFayette, CO 80026 (Ph. 303-926-0777). A Rainwater Basin checklist, based on surveys of the region's waterfowl production areas, is available on-line: http://www.rwbjv.org/pdf/RWBJV_bird_list.pdf.

A collective bird list for the Rainwater Basin and adjacent central Platte Valley, based on a survey by Lingle (1994), has more than 300 species, including 120 wetland species (35 breeders), and is available on-line: http://www.npwrc.usgs.gov/resource/birds/chekbird/r6/plandrwb.htm.

10. Valentine National Wildlife Refuge. Area 71,516 acres. Nebraska's largest national wildlife refuge, consisting mostly of Sandhills prairie, with sand dunes and intervening depressions that contain many shallow, sometimes lake-sized, marshes. Four grebes (eared, western, Clark's and pied-billed) nest here, as does the white-faced ibis, long-billed curlew, upland sandpiper, Wilson's phalarope and American avocet. The area and its birds have been described by Farrar (2004) and Johnsgard (1995, 2011). Up to 150,000 ducks can be found on the Refuge, with peak numbers occurring in May and October. The refuge checklist of 272 total species includes 100 wetland species (38 breeders) and is available on-line: http://www.npwrc.usgs.gov/resource/birds/chekbird/r6/valentin.htm. Located 22 miles south of Valentine, in Cherry County. Address: Hidden Timber Star Route, Valentine, NE 69201 (Ph. 402-376-378). URL: http://valentine.fws.gov/.

Kansas

1. Cheyenne Bottoms Wildlife Area. This famous state-owned wildlife area is about five miles north of Great Bend, in Barton County. It consists of about 18,000 acres of marshland, as well as adjacent bottomlands associated with the Arkansas River. The site is recognized as being of international importance for migratory shorebirds, at peak holding as many as 200,000 or more migrants. Notable breeding species include the least bittern, yellow-crowned night-heron, king rail, common gallinule and

snowy plover. A bird checklist of 325 species, including 145 wetland species, was published by Hoffman (1987) and also included by Zimmerman (1990) in his book on the area's ecology. The site's birding attractions were described by Zimmerman and Patti (1988), and also by Gress
and Janzen (2008). For information, contact Kansas Dept. of Wildlife &
Parks (Ph. 316-793-7730), URL: www.cheyennebottoms.net. The nearby
Kansas Wetland Education Center, managed by Fort Hayes State University (wetlandscenter.fhsu.edu) is a fine source of information on this and
other Kansas wetlands. Address: 592 NE K-156 Highway, Great Bend, KS
67530 (Ph. 620-786-7456).

2. Flint Hills National Wildlife Refuge. This refuge is on the upper end of
the John Redmond Reservoir of the Neosho River in Coffey County. It
includes 18,500 acres, most of which consists of the reservoir itself, and
is managed primarily for waterfowl. Notable breeding species include
wood duck, least bittern and upland sandpiper. During migration up to
100,000 waterfowl may be present. Many waterfowl overwinter here,
which attract substantial numbers of bald eagles. The site's birding aspects were described by Zimmerman and Patti (1988), Gress and Potts
(1993, and by Gress and Janzen (2008). A bird list of more than 290 species, including 113 wetland species (21 breeders), is available on-line:
http://www.npwrc.usgs.gov/resource/birds/chekbird/r6/flinthil.htm.
Address: P.O. Box 128, Hartford, KS 66854 (Ph. 316-392-5553).

3. Gardner Wetlands (Kansas City Power & Light Company Wetland Park).
This site is conveniently located near Kansas City, and notable for its attractiveness to migratory shorebirds and other wetland species. It was
described by Gress and Janzen (2008), and is a short distance west of
Gardner, in Johnson County. URL: www.gardnerkansas.govparks/park-
wetlands.php

4. Kirwin National Wildlife Refuge. This refuge is about ten miles southeast
of Phillipsburg, in Phillips County. It consists of 10,800 acres, mostly consisting of marshes, grasslands, croplands and a 5,000-acre reservoir impounded by the north fork of the Solomon River. Large numbers of ducks
(especially mallards) and Canada geese winter here. Migrating sandhill
cranes regularly stop here, and rarely whooping cranes are seen. The site's
birding opportunities were described by Zimmerman and Patti (1988), and
by Gress and Janzen (2008). A bird list of 234 total species, including 131
wetland species (10 breeders, mostly ducks and the least tern), is available

on-line: http://www.npwrc.usgs.gov/resource/birds/chekbird/r6/kirwin. htm. Address: Rte. 1. Box 103. Kirwin, KS (Ph. 913-543-6673).

5. Marais des Cygnes National Wildlife Refuge and Marais des Cygnes Wild-life Management Area. These two adjoining areas along the Missouri bor-der encompass about 15,000 acres of wetlands, prairie, deciduous wood-land, and transitional habitats. The region is an important wintering area and a major migration stopover point for both ducks and geese. Un-like the grassland-based wetlands of the region, Marais des Cygnes is a wooded riparian refuge, with several tree-nesting nesting herons, and attracts both waterthrushes and the swamp-nesting prothonotary war-bler. Both sites were described by Zimmerman and Patti (1988), and by Gress and Janzen (2008). Address of refuge: Rt. 2, Box 185A, Pleasan-ton, KS 68075. The refuge is administered by Flint Hills N.W.R. (Ph. 316-392-5553). For the state-owned wildlife management area, contact Kan-sas Dept. Wildlife & Parks (Ph. 913-351-8941). A list of 321 bird species, including 131 wetland species (22 breeders) is available on-line: http:// www.fws.gov/maraisdescygnes/Birding_Information.html.

6. McPherson Valley wetlands. This state-owned area of 1,310 acres is lo-cated northeast of Inman, in McPherson County. It includes Lake Inman, the largest natural lake in Kansas, as well as relict marshes that date back to the end of the Pleistocene about 10,000 years ago and that are impor-tant for migrating wetland birds. Birding opportunities there were de-scribed by Gress and Potts (1993). For more information contact Kansas Dept. Wildlife & Parks (Ph. 316-767-5900).

7. Neosho Wildlife Management Area. This state-owned area of 3,246 acres is the largest wetland in southeastern Kansas, and consists of wetlands and riparian woodlands in the Neosho River valley. It is especially impor-tant for migrating waterfowl and shorebirds, and its birds have been de-scribed by Zimmerman and Patti (1988), Gress and Potts (1993), and by Gress and Janzen (2005). Located one mile east of St. Paul, in Neosho County. For more information, contact Kansas Dept. Wildlife & Parks (Ph. 316-362-3671).

8. Perry Lake and Perry Lake State Park. This site in the Delaware River Valley of Jefferson County consists of Perry Lake (a reservoir of about 12,000 acres), marshes, mudflats, prairie, old fields and riparian woods. The area and its birds have been described by Zimmerman and Patti

(1988), Gress and Potts (1993), and Gress and Janzen (2005). Associated wetlands of interest here include Ferguson, Kyle and Lassiter marshes. Address of Perry Dam: U.S. Army Corps of Engineers, 10419 Perry Park Drive, Perry, KS 66073 (Ph. 785-597-5144). For information on Perry Lake State Park, contact the Kansas Dept. Wildlife & Parks (Ph. 913-246-3449) or the Kansas Division of Travel & Tourism (7995-537-43885).

9. Quivira National Wildlife Refuge. This outstanding marshland refuge is located 12 miles northeast of Stafford, Stafford County. It consists of 21,800 acres, including 4,700 acres of marsh, as well as grassland, farmlands, and sandhills. It is notable for its diverse wetland birds, especially sandhill and whooping cranes. As many as 500,000 geese, 100,000 ducks and 150,000 cranes have been observed here during migration. The refuge is nationally recognized as representing critical habitat for migrating whooping cranes, which typically pass through in April and late October to early November. Notable breeding species are 11 species of herons (including least bittern, great egret, snowy egret, yellow-crowned night-heron and white-faced ibis), four rails (black, king, sora and Virginia), plus the hooded merganser, eared grebe, snowy plover and black-necked stilt. The area and its birds have been described by Zimmerman and Patti (1988), Gress and Potts (1993) and Gress and Janzen (2005). A bird list of 340 total species, including 150 wetland species (52 breeding species, the largest number for any Kansas site), is available on-line: http://www.fws.gov/quivira/birdlist.htm. Address: R.R. 3, Box 48A, Stafford, KS 67578 (Ph. 316-486-2393). URL: www.fws.gov/quivira.

10. Tuttle Creek Lake and associated wetlands. This largest of Kansas reservoirs (about a miles wide and over 30 miles long) is about 15 miles northeast of Manhattan, It consists of 28,500 acres of public lands and wetlands associated with Tuttle Creek Lake. The reservoir and associated wetlands are important staging areas for migrating waterfowl and shorebirds. Wetlands with public access include Fancy Creek State Park, Olsburg Marsh (north of Olsburg on Shannon Creek Road), Carnahan Creek Park, Outlet Park, Tuttle Creek Cove Park, Stockdale Park, and River Pond State Park. The area and its birds have been described by Zimmerman and Patti (1988), Gress and Potts (1993) and Gress and Janzen (2005). Information on these sites can be obtained from the Visitor Center at Tuttle Creek Dam (Ph. 785-539-8511), the Manhattan Visitor's Bureau (Ph. 785-776-8829) or the Kansas Division of Travel & Tourism (7995-537-43885).

Breeding Species of Wetland Birds

In the following accounts, descriptive terms for relative abundance used here among species having very few state occurrences are my own. They do not exactly correspond with terms used by the Nebraska Ornithologists' Union or authorities for the other two states, as there are no general interstate agreements as to their meanings. "Accidental" or "accidental vagrant" here means that up to ten total reports (including undocumented ones) were known to me for a state, and "very rare" refers to those species with up to twenty state reports. Other terms that I have subjectively used to describe progressively more frequently encountered species are "rare," "occasional," "uncommon," "common" and "abundant".

"Casual" is not used here as an abundance descriptor, as it has had quite varied usages among different authorities. For example, Tallman, Swanson and Palmer (2002) used "accidental" mean up to two records of the species were obtained in the past ten years, and "casual" to mean 3-10 records had accrued in the prior ten years. Sharpe, Silcock and Jorgensen (2001) limited "accidental" to those species reported only once from the state, and "casual" to those species reported at least twice during a particular season. Thompson *et al* (2011) defined an "accidental" species as one far from its normal range and not likely to appear in the near future, a "vagrant" as a species to be expected every few years, and "casual" as a species occurring in small numbers most years. "Resident" here means that breeding in a state is known or assumed, whereas "visitor" implies that breeding is not known to occur there. Abbreviations in the text include N.W.R. (National Wildlife Refuge), W.P.A. (federal Waterfowl Production Area), and W.M.A. (state Wildlife Management Area). The term "Wildlife Area" is used in Kansas as an equivalent to Wildlife Management Area.

Sources of information relative to the three states are usually identified by citations, but in some cases information on breeding from other nearby states have been included without citations. Data on breeding in North Dakota are from Stewart (1975), and Iowa data are from Dinsmore *et al.* (1984). Oklahoma data are from Baumgartner and Baumgartner (1992).

35

Kansas egg records include some from Johnstone (1964) as well as from Thompson *et al.* (2011). National Breeding Bird Survey trend data are from Sauer *et al.* (2011). Waterfowl hunter-kill and population tends estimate are from the U. S. Fish & Wildlife Service (2009a. 2009b).

ORDER ANSERIFORMES – WATERFOWL

Family Anatidae – Ducks, Geese and Swans

Fulvous Whistling-Duck, *Dendrocygna bicolor*

A very rare migrant in Kansas, with records from six counties over 14 years, including a specimen from Kearny County. Accidental in South Dakota, with records from two counties (Tallman, Swanson and Palmer, 2002).

Breeding Status: Hypothetical. Nesting at Cheyenne Bottoms Wildlife Area was suspected in 1965 (Thompson *et al.*, 2011). An apparent nesting record is from Morton County, Kansas, in 1971 *(American Birds* 25:873).

Habitats: Typical original breeding habitat consisted of freshwater marshes with extensive beds of cattails and bulrushes. Recently the birds have colonized rice fields, particularly those heavily infested with weeds.

Nest Location: In freshwater marshes these birds typically construct their nests in clumps of living or dead bulrushes or in knotweeds, or they build floating nests in open water. Nests in rice fields are usually on levees, over water between levees, or attached to growing plants. On coastal marshes of Texas they typically are built over water 3-7 feet deep. No down is present in the nests; a likely reflection of dual incubation and adaptation to a warm climate.

Clutch Size and Incubation Period: From 10–16 eggs in nests of single females; the presence of supplemental eggs laid by other females ("dump nesting") often results in very large clutches. The eggs are white with a slightly roughened surface. The incubation period is 24-26 days. Single-brooded.

Time of Breeding: In Texas, egg dates range from May 10 to September 16. Downy young have been seen there as late as October 19, indicating a very long and rather irregularly timed breeding period.

Breeding Biology: Like other whistling ducks, this species is highly monogamous and probably forms lifelong pair bonds. Courtship displays are virtually nonexistent, at least as now understood. The best-known social displays are those associated with copulation, which occurs on water and is preceded by mutual head-dipping. A distinctive "step-dance" performed by both birds as they each lift one wing and rise side-by-side while treading water, follows mating. The female presumably builds the nest, but both sexes incubate. Incubation begins when the last egg is laid, and hatching is simultaneous. Both parents tend the young, which require about 65 days to fledge.

Comments: This tropically-distributed duck has one of the widest and most disjunctive distributions of all waterfowl, and is part of a seemingly generalized assemblage (*Dendrocygna* and *Thalassornis*) of duck species that are transitional in morphology and behavior between the goose-swan evolutionary complex and typical ducks.

Suggested Reading: Meanley and Meanley 1959; Johnsgard,1975, B.O.N.A. 562; Kear 2005.

Canada Goose, *Branta canadensis*

A common to abundant migrant throughout the three-state region, with widespread breeding and local overwintering. Canada geese have been raised and released widely, and now occurs throughout the region. Overwintering is increasingly common, especially among the larger races and more southern areas, where open water is available through the winter.

Migration: Forty-five initial spring sightings in Nebraska are from January 4 to April 3, with a median of March 27. Forty-one final spring sightings in Nebraska are from March 19 to May 30, with a median of April 28. Fifty-three initial fall sightings are from July 28 to December 20, with a median of October 13. Fifty-four final fall sightings are from October 18 to December 31, with a median of December 10. With the advent of warmer winters, a good deal of over-wintering occurs in the southern parts of the region, mostly at reservoirs and along ice-free rivers. A state-level analysis of four decade-long periods of Christmas Bird Counts (1967-68 to 2006–7) extending from North Dakota to the Texas panhandle indicated a late-December population peak in Kansas (Johnsgard and Shane, 2009).

Habitats: Migrant birds are found on large marshes, lakes or reservoirs, and nearby grain fields. Breeding is typical on prairie marshes, or sometimes on larger lakes with islands or muskrat houses.

Breeding Status: The giant race of the Canada goose (*B. c. maxima*) originally bred over much of the region concerned, south to central Kansas. Reintroductions at refuges and other localities have reestablished large Canada geese as breeding birds throughout the Great Plains.

Breeding Habitats: The historic breeding habitat of the giant Canada goose typically consisted of prairie marshes, especially those in the glaciated portions of the upper Great Plains. Some larger lakes were also used for breeding in earlier times, with the birds usually nesting on islands. Now Canada geese are prone to nest on farm ponds, in city parks, and other urban or suburban artificial wetlands.

Nest Location: Muskrat houses probably originally were important nest sites for this race of geese in prairie marshes, but emergent plants such as phragmites and bulrushes were no doubt frequently used. Where terrestrial predators are significant, islands are important nest locations. Nests are often some distance from water, such as in depressions in the prairie or under shrubs. Elevated nest sites are often used if available. Considerable down is normally present.

Clutch Size and Incubation Period: From 4–10 eggs, averaging about five. The eggs are white with a smooth surface. The incubation period is 26–29 days, averaging 28 days.

Time of Breeding: Nests in South Dakota have been reported from April 4 to May 28 (Peterson, 1995). In Nebraska, egg-laying probably begins in late March or early April. Egg dates in Kansas are from April 7 to May 5 (Thompson *et al.*, 2011). Young have been seen in Kansas as early as late April or early May.

Breeding Biology: Canada geese have strong, permanent pair bonds, and most begin to breed when two or three years old. Pair bonds are maintained by mutual displays, especially the "triumph ceremony". Males establish fairly large territories in marshes, usually including the same area and often the same nest site as in previous years, and unless nest sites are limited or predator pressures are present, the nests tend to be well scattered. The nest is constructed primarily by the female, with the· male standing guard and helping to some extent. Copulation occurs on the water, primarily during the egg-laying period, and incubation does not begin until the clutch is complete. Males remain close to the nest and take the major responsibility for guarding it but do not help incubate. Both sexes tend the young, which soon begin to fend for themselves. During the fledging period of about 70 days, both parents undergo a flightless molting period, and thereafter the family may leave the area, with the family bonds persisting through the winter.

Comments: The largest birds in the Great Plains region are of the locally breeding giant race *maxima*, weighing up to about 14 pounds (and historically even heavier), with the more westerly-breeding *moffitti* slightly smaller. Somewhat smaller still is *interior*, which breeds in southern Canada and commonly migrates through the region, especially in the east. Finally, the race *parvipes* is transitional in size between *hutchinsii* and *interior*, and migrates through the western parts of the three-state region. It is thought that gene exchange in central Canada between *parvipes* and *hutchinsii*, and farther west between *parvipes* and *minima,* is slight. Silcock (2006) evaluated 160 specimens collected in the 1880 by Dumont and Swenk (1934), and concluded that 70 belonged to the race *interior*, 70 to *parvipes*, 16 to *hutchinsii*, and three to *maxima,* as those races are currently defined (see a further discussion of the Great Plains races of the Canada goose under the account of the cackling goose later in this volume). In 2009 the population of the race *maxima* was estimated at 1.9 million birds. Canada goose populations have also been increasing. The average annual hunter-kill estimate for Canada geese in the U.S. during the five years 2004–8 totaled about 2.65 million birds, and kills have been progressively increasing since the 1960's. Estimated total annual Canadian kills from 1990–1998 for Canada geese ranged from about 183,000–274,000. All told, there may have been more than five million Canada geese in North America by 2009 (U.S.F.W.S., 2009a). Assuming that a total annual kill of nearly three million birds is correct, the recent (2011) fall North American population was likely to have been at least six or seven million birds. National Breeding Bird Surveys between 1966 and 2009 indicate that the species underwent a statistically significant population increase (9.8 percent annually) during that period, one of the highest rates of increase for any North American bird species.

  Suggested Reading: Brakhage 1965; Johnsgard 1975; B.O.N.A. 682; Kear 2005; Silcock, 2007.

Trumpeter Swan, *Cygnus buccinator*

  A rare to locally uncommon spring and fall migrant, and a local summer resident and winter visitor in the Sandhills of Nebraska and southwestern South Dakota. This species historically nested in the northern Great Plains, but was extirpated and apparently absent until the late 1960s, when recolonization occurred as a result of releases made at Lacreek N.W.R., South Dakota.

  Migration: Eight spring sightings in Nebraska are from January 24 to May 23, with a mean of March 28. Six fall sightings are from August 10 to November 7, with a mean of October 6. Local wintering occurs on various Sandhills rivers and creeks. Sightings of migrants along the eastern edge of the region may be the result of successful re-introductions into Minnesota and Iowa. A state-level analysis of four decade-long periods of Christmas Bird Counts (1967-68 to 2006–7) extending from North Dakota to the

Texas panhandle indicated a late-December population peak in South Dakota (Johnsgard and Shane, 2009).

Habitats: Migrants are found on lakes, large marshes, and impoundments. Breeding occurs on large shallow marshes or lakes having abundant submerged vegetation, emergent plants, and stable water levels. Wetlands used during the breeding season average about 180 acres, with about 75 percent open water and having slight to medium salinity levels.

Breeding Status: Since 1963 trumpeter swans have bred at Lacreek N.W.R. after having been introduced in 1960 from Red Rock Lakes N.W.R., Montana. They have since spread over much of western South Dakota, and have expanded into the Nebraska Sandhills (Ducey, 1999). Nesting has since occurred in many of Nebraska's Sandhills lakes, especially in Cherry and Grant counties, and has also been reported from marshes in Arthur, Brown, Garden, McPherson and Sheridan counties. There were seven confirmed nestings during the 1984–1989 atlasing period in Nebraska (Mollhoff, 2001) and three in South Dakota from 1988 to 1993 (Peterson, 1995). By 1987 the South Dakota population had reached at least 268 birds, and by 1995 the Nebraska population had reached about 150 birds. There were possibly as many as 5,000 trumpeter swans in the expanding interior North American population by 2010, which is located in widely scattered restoration sites from South Dakota east to Ontario. Restoration in Minnesota has been notably successful, and birds from this flock probably account for at least some of the individuals seen in eastern parts of the region during migration periods.

Breeding Habitats: Typical breeding habitat consists of large, shallow marshes to shallow lakes, with an abundance of submerged plants and emergent vegetation, and stable water levels. The emergent plants provide important nesting cover, and the submerged vegetation is the major food source.

Nest Location: Nests are greatly scattered, owing to extreme territorial behavior of adults, and nest sites are usually used for several years. Island locations are preferred over shoreline sites, and when nests are built in emergent vegetation the water is usually 12–36 inches deep. Sometimes muskrat houses or beaver lodges serve as nest sites.

Clutch Size and Incubation Period: From 3–9 eggs, averaging about five. The eggs are creamy white and somewhat granular. The incubation period is 32–37 days, usually about 34 days. Single-brooded.

Time of Breeding: In South Dakota, the Lacreek N.W.R. records indicate that nest-building occurs from April 3 to about May 20 and hatching from May 20 to July 1. Nests in South Dakota have been reported from April 15 to June 12 (Peterson, 1995). Fledging occurs from September 20 to October 16. Incubating adults have been seen in late May and early June.

Breeding Biology: Trumpeter swans pair for life, and each pair returns to its nesting area in spring as soon as the weather allows. Territories are established that average more than 30 acres, sometimes more than 100 acres, and are vigorously defended; the adults even exclude their own offspring of previous years. The male initiates territorial defense, but the female participates in mutual "triumph ceremonies" after territorial disputes and also helps defend the nest site. Both sexes help construct the rather bulky nest, which may require a week or more. The eggs are laid at two-day intervals, and no incubation is performed until the clutch is complete. Thereafter the female normally performs all the incubation, while the male defends the nest. Most of the cygnets hatch within a few hours of each other and are led from the nest within 24 hours of hatching. The nest may later be used for resting or brooding, but often the brood is led some distance from the nest for rearing on quiet and secluded ponds. The fledging period is approximately 100 days, which occupies the entire summer and makes it impossible for birds to renest after nest failure.

Comments: This is the heaviest of all of North American wetland birds, with records of birds weighing as much as 36 pounds. These swans prefer to nest on marshes having a substantial amount of freedom from human disturbance, and are highly territorial, thus requiring large wetlands for breeding. The birds are only slightly migratory, sometimes moving south as far as Kansas during winter, so the Nebraska Sandhills are one of the most important Great Plains wintering areas for the species.

Suggested Reading: Banko 1960; Johnsgard 1975; B.O.N.A. 105; Kear 2005.

Wood Duck, *Aix sponsa*

An uncommon spring and fall migrant and summer resident in eastern parts of the three-state region, but less common westward, and infrequent in western Kansas, the Nebraska Panhandle and Sandhills. Breeding in South Dakota has been confirmed west to the Black Hills, and in Kansas west to the counties bordering Colorado. Most migrants are found in eastern parts of the region.

Migration: Sixty-nine initial spring sightings in Nebraska are from January 17 to June 7, with a median of March 28. Half of the sightings fall within the period March 13 to April 8. Thirty-five final fall sightings are from September 10 to December 31, with a median of October 21. Half of the fall records fall within the period October 3–30. A state-level analysis of four decade-long periods of Christmas Bird Counts (1967-68 through 2006–7),

extending from North Dakota to the Texas panhandle, indicated a late-December population peak in Oklahoma (Johnsgard and Shane, 2009).

Habitats: Throughout the year this species is associated with tree-lined rivers, creeks, oxbows and lakes, and usually breeds near slow-moving rivers, sloughs or ponds where large trees are found.

Breeding Status: Largely restricted to wooded rivers east of the 100th meridian, but increasingly occurring locally west to central South Dakota and the eastern thirds of Nebraska and Kansas. Nebraska breeding was limited in the early 1900's to the Missouri's forested valley and the lower portions of the Platte Valley, probably west to about Kearney. This species has significantly extended its range westward across the Great Plains, and in Nebraska probable family groups have been seen as far west as Dawes, Garden and Scotts Bluff counties. South Dakota breeding records extend to some westernmost counties (Butte and Pennington counties), and to western Kansas (Hamilton and Cheyenne counties). There were 46 confirmed nestings during breeding bird atlas surveys in South Dakota (Peterson, 1995), 84 in Nebraska (Mollhoff, 2001), and 73 in Kansas (Busby & Zimmerman, 2001).

Breeding Habitats: The wood duck breeds in floodplain forests along rivers, creeks, and oxbows and around wooded lakes but is generally associated with slow-moving rivers, sloughs, or ponds where large trees are found. Forests providing acorns or other large seeds are desirable, and wetlands with an abundance of flooded shrubs or trees and depths no greater than 18 inches are especially favored.

Nest Location: Nests are in natural or artificial cavities. Favored trees are at least 16 inches in diameter, having openings at least 3 ½ inches wide and interior cavities at least eight inches in diameter. The cavity must be well drained and the entrance well protected from the weather. High cavities with small entrances are preferred, as are locations over water. Likewise, trees growing in clusters or groves are favored over isolated trees, and open stands of trees are preferred over dense stands.

Clutch Size and Incubation Period: From 12–16 eggs, averaging about 14 in clutches produced by a single female, often more in "dump nests" produced by several females. The eggs are creamy white with a smooth surface. The incubation period is 25–37 days, averaging 30 days. Normally single-brooded.

Time of Breeding: Egg dates in North Dakota are from May 7 to June 14, and dates of dependent young are from June 3 to September 8. Nests in South Dakota have been reported from April 19 to June 29 (Peterson, 1995). Young in Nebraska have been seen in late May. Kansas egg dates are from March 21 to May 10, with mid-April a probable peak of egg-laying.

Breeding Biology: Pair bonds are established each year, after a prolonged period of courtship displays. No definite territorial behavior exists, but males assist females in seeking out suitable nest sites, which may take days. Competition for nest sites is frequent, and as a result collective "dump nests" produced by two or more females are locally prevalent. The female does the incubation, and males normally desert their mates before hatching. The female raises the brood, which fledges at about 60 days of age. Renesting after loss of the first clutch is fairly frequent, and a second brood may be raised on rare occasions.

Comments: Wood ducks have become much more widespread and common in the region recently, both because of nest-box erection programs and also the increasing growth and maturation of riverine forests along major river systems. National Breeding Bird Surveys between 1966 and 2009 indicate that the species underwent a statistically significant population increase (2.4 percent annually) during that period.

Suggested Reading: Grice and Rogers 1965; Johnsgard 1975; B.O.N.A. 169; Kear 2005.

Gadwall, *Anas strepera*

A common to abundant spring and fall migrant and a common summer resident in the three-state region,

Migration: The range of 48 initial spring sightings in Nebraska is from January 3 to June 8, with a median of March 28. Half of the records fall within the period March 6 to April 8. Fifty final fall sightings range from October 4 to December 31, with a median of November 21. Half of the records fall within the period November 2 to December 2. A state-level analysis of four decade-long periods of Christmas Bird Counts (1967-68 to 2006–7) extending from North Dakota to the Texas panhandle indicated a late-December population peak in Oklahoma (Johnsgard and Shane, 2009).

Habitats: Migrants are normally found in shallow marshes and sloughs, and sometimes on deeper waters such as lakes and reservoirs. Nesting occurs preferentially on shallow prairie marshes, especially those having grassy or weedy islands or surrounding weedy cover.

Breeding Status: Breeds over nearly the entire northern half of the region, being relatively common in South Dakota, as well as in the Nebraska Sandhills. There were 34 confirmed nestings during breeding bird atlas surveys in South Dakota (Peterson, 1995), eight in Nebraska (Mollhoff, 2001), and ten in Kansas (Busby & Zimmerman, 2001). In Nebraska, it primarily breeds north of the Platte River and especially in the Sandhills. It also nests south to the Rainwater Basin. Although it is a regular breeder at Cheyenne Bottoms Wildlife Area and Quivira N.W.R., there is only one other definite breeding locality (Slate Creek wetlands, Sumner County) for Kansas. However, it is a historic breeding species in Ellis, Grant, Linn, Meade, Russell, Trego and Wilson counties (Thompson *et al.*, 2011).

Breeding Habitats: Breeding occurs on a variety of mostly temporary or semi permanent wetlands, ranging from fresh to sub-saline. Shallow prairie marshes that are relatively alkaline are apparently preferred over deeper, more permanent marshes, and those with grassy or weedy islands are also heavily used.

Nest Location: Nests are built on dry ground under a variety of cover types, in particular amid broad-leaved weeds. Dry upland sites are preferred to wetter areas, and dense cover is preferred to sparser cover. Vegetation 1–3 feet in height, especially on islands, is frequently used for nesting; island nesting in dense populations is at times almost colonial. The nest cavity is lined with rather dark grayish down.

Clutch Size and Incubation Period: From 7–13 eggs (667 nests in North Dakota averaged 9.9). The eggs are dull creamy white. The incubation period is 25–27 days, averaging 26 days. Renesting is fairly common, and such nests have a slightly smaller average clutch size than initial nesting efforts.

Time of Breeding: Nests in South Dakota have been reported from May 2 to June 22 (Peterson, 1995). Egg-laying in Nebraska probably extends into early June (Mollhoff, 2001). Egg dates in Kansas are from June 12–July 10) Thompson *et al.*, 2011).

Breeding Biology: Gadwalls form their pair bonds relatively early, during a period of social courtship involving aquatic display as well as aerial

chases. Most birds are paired by the time they arrive. Once on their nesting grounds, pairs establish home ranges that may exceed 50 acres, often overlapping with the home ranges of other pairs. Territorial behavior as such is not significant, and nests are often close together, especially on islands. The female constructs the nest alone and is usually abandoned by her mate about a week or two after incubation has begun. The hen thus raises her brood alone, usually on deepwater marshes unlikely to dry up before fledging, which requires 7–8 weeks.

Comments: North American breeding grounds surveys of gadwalls in 2009 indicated a total population of 3.05 million birds, 71 percent above the long-term average (U.S.F.W.S., 2009a). The average annual hunter-kill estimate in the U.S. during the five years 2004–8 has been about 1.46 million birds, and estimates have exhibited a long-term increase since the 1960's. Estimated total annual Canadian kills from 1990–1998 ranged from about 32,000–50,000. National Breeding Bird Surveys between 1966 and 2009 indicate that the species underwent a statistically significant population increase (2.3 percent annually) during that period.

Suggested Reading: Oring 1969; Johnsgard 1975; B.O.N.A. 283; Kear 2005.

American Wigeon, *Anas americana*

A common to locally abundant spring and fall migrant throughout the region. It a uncommon breeder in South Dakota, and in Nebraska a local and generally uncommon breeder. It is apparently mostly confined to the northwestern parts of the Sandhills (south and east to Garden and Holt counties). It is an extremely rare nester in Kansas.

Migration: Sixty-seven initial spring sightings in Nebraska range from January 9 to May 28, with a median of March 22. Half of the sightings fall within the period March 6 to March 30. Thirty-four final spring sightings in Nebraska are from March 27 to June 6, with a median of May 3. Fifty initial fall sightings are from August 28 to December 17, with a median of September 30. Fifty final fall sightings are from October 9 to December 31, with a median of November 18. A state-level analysis of four decade-long periods of Christmas Bird Counts (1967-68 to 2006–7) extending from North Dakota to the Texas panhandle indicated a late-December population peak in northwestern Texas (Johnsgard and Shane, 2009).

Habitats: During migration these birds are sometimes found on large lakes or reservoirs, but forage where submerged plants can easily be reached from the surface or around the shoreline in grassy meadows. Breeding is usually done on marshes or lakes with abundant aquatic food at or near the surface, and especially those with adjacent sedge meadows or brushy, partially wooded habitats nearby.

Breeding Status: A common to uncommon breeder in the Nebraska Sandhills, but there are no specific nesting records for the more southerly parts of this region. It nested at Cheyenne Bottoms Wildlife Area, Barton County, Kansas, in 1963, and in Seward County in 1982 (Thomson *et al.*, 2011). There were four confirmed nestings during breeding bird atlas surveys in South Dakota (Peterson, 1995), one in Nebraska (Mollhoff, 2001), and two in Kansas (Busby & Zimmerman, 2001).

Breeding Habitats: American wigeons favor marshes or lakes with abundant aquatic food at or near the surface, but with limited emergent aquatic vegetation. Areas surrounded by sedge meadows are favored, as are those with partly wooded or brushy habitats near the water.

Nest Location: Nests are on dry land, often 100 yards or more from water. The surrounding cover is often of sedges, rushes, mixed prairie grasses, or weeds, but the nests are also sometimes placed near the base of a tree. The nest is simply a slight depression in the soil, well lined with light grayish down.

Clutch Size and Incubation Period: From 712 eggs, averaging eight or nine in most areas. The eggs are creamy white with a smooth surface. The

incubation period is 23–24 days. Single-brooded, but renesting apparently is frequent.

Time of Breeding: North Dakota egg dates range from May 31 to July 13, and dates of dependent young are from June 26 to September 21. Nests in South Dakota have been reported from May 20 to June 20 (Peterson, 1995). Egg-laying in Nebraska probably occurs in May and June.

Breeding Biology: Wigeons form seasonally monogamous pair bonds after a period of social courtship in winter and spring. Males perform fairly simple displays, mainly involving calling, chin-lifting, and raising the folded wings high above the back. After pair-formation, pairs establish a home range on marshes ranging from less than an acre to more than 20 acres in area. There is no territorial defense, although males evict other males from the vicinity of their mates. Nest sites are well hidden, and shortly after incubation begins males abandon their mates. The female thus incubates and rears the brood alone. Broods are reared on relatively open marshes, and fledging occurs at about 70 days of age.

Comments: This species often associates with gadwalls, but is more inclined to feed on grassy vegetation along shorelines than are other surface-feeding ducks. Gadwalls and wigeons usually arrive about the same time as green-winged teal in spring, comprising a "second wave" of birds that seasonally appear shortly after mallards and pintails. North American breeding grounds surveys in 2009 indicated a total population of 2.47 million birds, five percent below the long-term average (U.S.F.W.S., 2009a). Total U.S. kills have averaged about 796,000 during the five years 2004–8, with no clear long-term directional trend. Estimated total annual Canadian kills from 1990–1998 ranged from about 37,000–51,000. National Breeding Bird Surveys between 1966 and 2009 indicate that the species underwent a statistically significant population decline (3.4 percent annually) during that period.

Suggested Reading: Johnsgard 1975; Sowls 1978; B.O.N.A. 401; Kear 2005.

American Black Duck, *Anas rubripes*

A rare migrant in the eastern half of the region, very rare in the west. In Nebraska, it has been observed west to Dawson, Keith and Cherry counties. In South Dakota and Kansas it is a rare migrant in the east, and a very rare breeder.

Migration: Nine spring Nebraska records range from March 1 to May 26, with a mean of March 26. There are fall Nebraska records from mid-September to December 22, and the species has been captured during winter banding operations in eastern Nebraska.

Habitats: Throughout nearly all of their range black ducks are associated with coastal marshes and eastern forests, and forest seems to represent their primary habitat. In the interior, the birds are found on fairly alkaline

marshes, acidic bogs and muskegs, stream margins, and lakes and ponds, especially those near woodlands.

Breeding Status: There are records of breeding at Waubay N.W.R., Day County, South Dakota until the 1950's and at Sand Lake N.W.R, Brown County, in 1981 and 1996 (Tallman, Swanson and Palmer, 2002). There is a single Kansas breeding record from Cheyenne Bottoms Wildlife Area in 1969 (Thompson *et al.,* 2011).

Nest Location: The margins of woody areas appear to be favored for nesting, with more open areas such as marshes or cultivated fields a secondary preference. Plants that serve as cover for the nest frequently are shrubby forms with evergreen or persistent leaves and dense branching patterns that provide excellent overhead concealment. A dry nest substrate, such as dead leaf litter, is an important component.

Clutch Size and Incubation Period: From 6–12 eggs (average of various studies, about nine). The eggs vary from creamy white to pale greenish or buffy. The incubation period is 26–27 days, beginning with the last egg. Single-brooded, but replacement clutches are frequent.

Time of Breeding: This rare species bred in Kansas at Cheyenne Bottoms Wildlife Area in 1969, when a nest with three eggs was found on July 7 (Thompson *et al.,* 2011). Nests with eggs or adults with young have been seen in Iowa during June (Dinsmore, 1984).

Breeding Biology: Apart from their ecological preferences, there are virtually no differences in the breeding biology of the black duck and the common mallard. Their generally complementary ranges reduce competition between them, but in recent years the extent of range overlap has increased and hybridization between the two types has become more prevalent. There is no good evidence that the black duck is extending its breeding range into the Great Plains, and it will probably continue to be a very rare breeder there.

Comments: In Nebraska, black ducks are seen only infrequently among mallard flocks, and many of these birds are actually mallard × black duck hybrids. The 2009 U.S. winter surveys of this species indicated a population of about 210,000 birds, whereas recent breeding surveys suggest that about 500,000 birds might be present (U.S.F.W.S., 2009a). Rose & Scott (1997) suggested 1990's populations of about 210,000 for the Atlantic flyway, and 90,000 for the Mississippi flyway. Even more of the black duck's original range has been impacted by competition from and hybridization with northern mallards than was the case during the 1970's. Most evidence indicates that the species has been in a long-term population decline in eastern North America, especially relative to mallards in the same region. Hunter-kill estimates of black ducks in the Atlantic flyway have recently dropped to about one-third of those occurring in the late 1960's (90,000 in 2008), with an average nationwide estimate of about 125,000 for the years 2004–8. However, kills of black duck × mallard hybrids have exhibited a slight increase, with a long-term average of about 8,000 hybrids taken in the Atlantic flyway during 2008, or nearly ten percent of total recent annual kills for the black duck in that flyway. This estimate of hybrid frequencies is more than three times higher than those I summarized for the 1960's, which included an estimate of 2.7 percent hybrids relative to the overall Atlantic flyway black duck population (Johnsgard, 1961, 1967). Nationwide, average kills of hybrids have been about 14,300 annually during the five years 2004–8, which also represent about ten percent of the combined black duck–hybrid sample. Estimated total annual Canadian kills from 1990–1998 ranged from about 153,000–243,000, or about twice total recent U.S. kill estimates. National Breeding Bird Surveys between 1966 and 2004 indicate that the species underwent a statistically significant population decline (0.8 percent annually) during that period.

Suggested Reading: Coulter and Miller 1968; Johnsgard 1975; B.O.N.A. 481; Kear 2005.

Mallard, *Anas platyrhynchos*

An abundant migrant and a locally common summer resident throughout the region. Wintering birds are common wherever open water occurs. Breeding occurs on wetlands in all areas.

Migration: Forty-three initial spring sighting are from January 1 to May 29, with a median of March 12. Half of the records fall within the period March 2–April 3. Sixty-four final fall sightings are from August 25 to December 31, with a median of November 27. Half of the sightings fall within the period November 21–December 28. A state-level analysis of four decade-long periods of Christmas Bird Counts (1967-68 to 2006–7) extending from North Dakota to the Texas panhandle indicated a late-December population peak in Kansas (Johnsgard and Shane, 2009).

Habitats: Breeding birds favor fairly shallow waters, either still or slowly flowing, and surrounding dry areas of non-forested vegetation. Migrants are often found on large marshes, lakes or reservoirs, especially where nearby grain fields provide food.

Breeding Status: Breeds over the entire region. There were 99 confirmed nestings during breeding bird atlas surveys in South Dakota (Peterson, 1995), 63 in Nebraska (Mollhoff, 2001), and 314 in Kansas (Busby & Zimmerman, 2001).

Breeding Habitats: One of the most widespread and adaptable species of ducks, the common mallard occupies a diversity of water types and surrounding environments. Fairly shallow waters, either still or slowly flowing, and surrounding dry sites of non-forested vegetation seem to be its preferred breeding habitat. Mallards will sometimes breed in forested areas, but never in large numbers.

Nest Location: Nests are on dry ground, usually under relatively tall grass or herbaceous vegetation, and are generally well concealed from above and from all sides. Grasses 1–4 feet tall seem to be the most common nest cover, but weeds such as thistles and nettles are also frequently used. The nest is a shallow depression in the soil, well lined with brown down.

Clutch Size and Incubation Period: From 5–15 eggs (118 North Dakota nests averaged 9.6). The eggs are creamy white to greenish white with a smooth surface. The incubation period is 24–30 days, averaging 28 days. Normally single-brooded, but renesting is regular after nest loss, and a few cases of double-brooding have been reported.

Time of Breeding: Nests in South Dakota have been reported from April 25 to July 6 (Peterson, 1995). Young in Nebraska are usually seen by the first week in June (Mollhoff, 2001). Kansas egg records are for the period April 1 to June 10, with egg-laying most frequent during the first ten days of May.

Breeding Biology: Mallards begin social display early in the fall, with many adults probably forming new pair bonds with earlier mates, and those hatched the previous summer beginning courtship for the first time. By spring, nearly all females have formed pair bonds, and on arrival at their breeding grounds pairs spread out across the available habitat. Home ranges of such pairs vary greatly in size but at times may exceed 700 acres; spacing is enhanced by males' evicting other males from the vicinity of their mates. Females choose their nest sites and are abandoned by their mates when incubation gets under way. The newly hatched young are quickly led to water, and the fledging period is 55–59 days in Manitoba. Mallards often try to renest if their first attempt fails; the clutch sizes of renesting efforts tend to be slightly smaller than the original clutches.

Comments: This is the region's commonest duck, and probably the most popular species among hunters. It is a hardy bird, usually over-wintering in large numbers, and breeding even in locations as unlikely as the heart of Lincoln and Omaha wherever urban lakes and streams allow. The 2009 North American breeding population was recently estimated at 8.5 million birds, 13 percent above the long-term average (U.S.F.W.S., 2009a). The average annual hunter-kill estimate in the U.S. during the five years 2004–8 has been about 4.62 million birds, with no clear directional long-term trend. National Breeding Bird Surveys between 1966 and 2009 indicate that the species had a statistically significant population decline (0.2 percent annually) during that period.

Suggested Reading: Girard 1941; Johnsgard 1975; B.O.N.A. 658; Kear 2005.

Mottled Duck, *Anas fulvigula*

A rare summer visitor in southern Kansas, and a breeder at Cheyenne Bottoms Wildlife Area during the 1960's and 1970's. A possible mottled duck was taken in Rock County, Nebraska, in October of 1969 (*Nebraska Bird Review* 38:80). A 1921 Nebraska specimen has been proven to be a hybrid (*N. Mex. Dept. Game and Fish Bulletin* 16, 1977).

Breeding Status: There are breeding records for Cheyenne Bottoms Wildlife Area, Barton County, Kansas, from 1963 to 1977.

Habitats: The preferred location is much like that of the common mallard, but the species inhabits brackish and saltwater habitats. Coastal marshes with extensive emergent vegetation are the primary breeding habitat, but the birds also breed in coastal prairies, bluestem meadows, and fallow rice fields.

Nest Location: Grass cover, usually fairly tall and dense, seems to be the preferred location for these coastal-nesting mallards. Distance from water is variable, and probably depends on local topography. The down lining of the nest is brown.

Clutch Size and Incubation Period: From 5–15 creamy white to greenish white eggs. The incubation period is 25–27 days. Probably single-brooded, but up to five nesting attempts have been reported in a single female.

Time of Breeding: Kansas egg records are for the period June 17–27. In Texas, eggs have been reported from March 18 to July 21, and dependent young have been seen from April 3 to August, except for one remarkable case of a brood observed in December.

Breeding Behavior: In most respects the behavior and breeding biology of this species is mallard-like. But it is possible that the pair bond may be relatively continuous in this population, even though the male does not seem to be present while the brood is reared. The fledging period of 54–60 days is virtually identical to that of mallards, and during this period the male apparently migrates south to the coastal areas of Texas to molt.

Comments: The mottled and Florida ducks' populations might consist of about 56,000 birds in Florida (Florida duck) and 500,000–800,000 in Texas and Louisiana (mottled duck) **(B.O.N.A. 81),** The average annual hunter-kill estimate in the U.S. of combined mottled and Florida ducks during the five years 2004–8 has been about 70,000 birds. Although the kill estimates have remained fairly steady recently, they have undergone a gradual long-term decline since the 1960s.

Suggested Reading: Engeling 1950; Johnsgard 1975; B.O.N.A. 81; Kear 2005.

Blue-winged Teal, *Anas discors*

An abundant spring and fall migrant and common summer resident throughout the three-state region. Breeding is regular in all areas.

Migration: Sixty-eight initial spring sightings in Nebraska range from February 10 to June 1, with a median of April 2. Half of the sightings fall within the period March 28 to April 10. Eighty-eight final fall sightings are from August 19 to December 31, with a median of October 10. Half of the records fall within the period September 24 to October 23.

Habitats: Migrants are found on generally shallow ponds, ditches, marshes, and the like, and rarely occur in deep open water. Breeding is typically in marshes surrounded by native prairies and grassy sedge meadows.

Breeding Status: Breeds commonly over the region, and is the commonest surface-feeding duck breeder in much of it. There were 56 confirmed nestings during breeding bird atlas surveys in South Dakota (Peterson, 1995), 41 in Nebraska (Mollhoff, 2001), and 19 in Kansas (Busby & Zimmerman, 2001).

Breeding Habitats: The highest concentrations of blue-winged teal are in marshes surrounded by native prairies, especially the tallgrass prairies. Relatively small, shallow ponds or marshes are favored over larger and deeper ones during breeding, especially where grassy or sedge meadows are nearby.

Nest Location: Nests are on dry land, often very near water, particularly in sedges or grasses that average about a foot high. Steep slopes and very

dense cover are avoided, and nests are often placed about halfway between water and the highest surrounding point of land in gently rolling country. The nests are well lined with grasses and dark down having conspicuous white centers.

Clutch Size and Incubation Period: From 8–13 eggs (349 nests in North Dakota averaged 10.3). The eggs are creamy white to pale olive-white with a slightly glossy surface. The incubation period averages about 24 days, starting with the last egg laid. Single-brooded, but renesting efforts are frequent.

Time of Breeding: Nests in South Dakota have been reported from May 18 to July 8 (Peterson, 1995). In Nebraska, young birds have been seen as early as June 2 (Mollhoff, 2001). Kansas egg dates are from May 1 to July 18.

Breeding Biology: Pair bonds are formed fairly late in blue-winged teal, mainly during the migration northward, but some displays may occur on the nesting grounds. Pairs are relatively tolerant of other pairs and often center their home ranges on very small ponds or even roadside ditches. The female chooses the nest site and builds the nest, while the male waits nearby. After incubation begins the pair bond is dissolved, and males often fly elsewhere to complete their summer molt. Females take their broods to water after hatching and usually raise them in rather heavy brooding cover. The fledging period is about six weeks, and females also begin to molt at about the time the young are fledged.

Comments: This is the commonest migrant "teal" in the state, and one of the latest duck species to arrive in spring, owing to its long migration from wintering grounds sometimes as far away as northern South America. It is also probably the most common breeding duck in Nebraska, but few remain long in the fall, the birds usually departing shortly after the first freezing weather. North American breeding grounds surveys in 2009 indicated a total population of 7.4 million birds, 60 percent above the long-term average (U.S.F.W.S., 2009a). The average annual hunter-kill estimate in the U.S. for combined blue-winged and cinnamon teal during the five years 2004–8 has been about 870,000 birds, but annual estimates been quite variable, and may reflect the influence of special early-season hunting periods for teal. Estimated total annual Canadian kills from 1990–1998 ranged from about 22,000–53,000. National Breeding Bird Surveys between 1966 and 2009 indicate that the species underwent a statistically non-significant population decline (0.1 percent annually) during that period.

Suggested Reading: Dane 1966; Johnsgard 1975; B.O.N.A. 81; Kear 2005.

Cinnamon Teal, *Anas cyanoptera*

An uncommon spring and fall migrant in the western half of the three-state region, becoming rarer eastwardly but observed east to counties along the Missouri River in Nebraska. Probably a local summer resident in western parts of the entire region.

Migration: Sixty-two initial spring sightings in Nebraska are from January 9 to June 6, with a median of April 26. Half of the sightings fall within the period April 8 to May 10. Six fall Nebraska records are from July 13 to November 14, with a mean of September 19.

Habitats: This species occupies the same shallow marshy habitats as does the blue-winged teal, but favors more alkaline waters where they are available.

Breeding Status: Apparently a rare breeding species at the eastern limit of its range in our region, but confusion with the blue-winged teal makes the status of this species difficult to ascertain. There were several probable breeding records during breeding bird atlas surveys in South Dakota (Peterson, 1995) and Kansas (Busby & Zimmerman, 2003). Probable broods have been seen at Lacreek N.W.R. in South Dakota and a nest is reported from Day County (Tallman, Swanson and Palmer, 2002). In Nebraska, cinnamon teal are regularly present during summer at Crescent Lake N.W.R., and there is at least one breeding record for Garden County. They definitely bred at Cheyenne Bottoms Wildlife Area, Barton County, Kansas, in 1969 (Thompson *et al.*, 2011).

Nest Location: This species usually nests in fairly low herbaceous cover 12–15 inches tall, often consisting of grasses, sedges, or broadleaf weeds. Such cover is favored for nesting if it provides excellent concealment and is close to stands of taller vegetation. Islands that provide low grasses are also preferred. The nest is lined with down nearly identical to that of blue-winged teal.

Clutch Size and Incubation Period: From 8–13 eggs, averaging about nine. The eggs are white to pale pinkish buff with a slight gloss. The incubation period is 24–25 days, starting with the last egg. Single-brooded, but a persistent renester.

Time of Breeding: Probably like that of the blue-winged teal. In Nebraska, a nest with eggs was seen from May 24 until early June (Mollhoff, 2001).

Breeding Biology: The social behavior and breeding biology of the cinnamon teal are extremely similar to those of the blue-winged teal, and in a few areas they breed on the same marshes, nesting at the same time and using the same habitats. Nesting densities of cinnamon teal in the middle of their range are appreciably higher than those of blue-winged teal, however, and their home ranges tend to be very small.

Comments: Persons wanting to see this beautiful teal should consider visiting Crescent Lake N.W.R. in June, when as many as 6–8 males might be seen on a good day. Females are almost impossible to distinguish from those of blue-winged teal, but have somewhat longer and wider bills, and are generally more uniformly brownish. Wild hybrids are occasionally reported. The North American cinnamon teal population has been estimated as 260,000 birds (Wetlands International, 2002). Hunter-kill figures for this species are not available, since they are combined with those of blue-winged teal. National Breeding Bird Surveys between 1966 and 2009 indicate that the species underwent a statistically significant population decline (0.8 percent annually) during that period.

Suggested Reading: Spencer 1953; Johnsgard 1975; B.O.N.A. 209; Kear 2005.

Northern Shoveler, *Anas clypeata*

A common to abundant spring and fall migrant, and a common to uncommon summer resident in the three-state region, with breeding most frequent in the Sandhills area, and decreasing southeastwardly. It is a migrant throughout the Plains States, and breeds locally except in the southern and southeastern portions.

Migration: Seventy initial spring sightings in Nebraska are from January 27 to June 6, with a median of March 23. Half of the sightings fall within the period March 11–30. Sixty-two final fall sightings range from September 5 to December 31, with a median of November 4. Half of the records fall within the period October 20 to November 20. A state-level analysis of four decade-long periods of Christmas Bird Counts (1967-68 to 2006–7) extending from North Dakota to the Texas panhandle indicated a late-December population peak in northwestern Texas (Johnsgard and Shane, 2009).

Habitats: Migrants utilize aquatic habitats rich in zooplankton and phytoplankton, and during the nesting season the birds favor shallow prairie marshes rich in those food sources. Non-wooded shorelines are preferred over wooded ones, and mud-bottom ponds are also apparently preferentially used.

Breeding Status: Breeds commonly in South Dakota, is common to uncommon in Nebraska, and occasionally nests in Kansas (Atchison, Barton, Finney, Kearny and Seward counties (Thompson *et al.*, 2011). There were 16 confirmed nestings during breeding bird atlas surveys in South Dakota (Peterson, 1995), nine in Nebraska (Mollhoff, 2001), and four in Kansas (Busby & Zimmerman, 2001).

Breeding Habitats: Shallow prairie marshes with an abundance of plant and small animal life floating on the surface provide ideal shoveler habitat. Submerged plants whose leaves reach or nearly reach the surface, such as pondweeds, also provide food by supporting an abundant aquatic invertebrate life. Non-wooded shorelines are preferred to wooded ones, and muddy Bottoms seem to be preferred.

Nest Location: Shovelers usually build their nests well away from water, in grassy cover that is less than a foot tall and almost never more than two feet tall. Broad-leaved weeds and shrubby cover are used secondarily for nesting. The nest is a shallow depression lined with some vegetation and with brownish down having lighter centers.

Clutch Size and Incubation Period: From 8–13 eggs (54 North Dakota clutches averaged 10.2). The eggs are buffy, usually with a greenish tint. The incubation period is 22–25 days. Single-brooded, but renesting efforts are frequent.

Time of Nesting: North Dakota egg dates range from May 6 to July 20, and records of dependent young range from June 8 to September 17. Nests in South Dakota have been reported from April 29 to June 12 (Peterson, 1995), and in Nebraska nests with eggs have been seen in early to mid-June.

Breeding Biology: Shovelers begin pair-formation on their wintering grounds and continue it through their arrival on the breeding grounds. Most of the displays are aquatic, but there are also "jump-flights" and aerial chases associated with courtship. The birds are seasonally monogamous, and at least in captivity some birds re-mate with previous mates while others choose new ones. The pairs spread out over the breeding habitat and have been described as territorial by some workers, while others have simply reported that they occupy overlapping home ranges from 15 to 90 acres in area. The females do all the incubation, and the males abandon them during the incubation period. The fledging period is about 6–7 weeks.

Comments: Although generally despised by hunters because of their reputed poor taste and oversized bill, the shoveler's bill is a marvelously adapted structure, allowing the birds to extract plankton-sized materials from water. Shovelers arrive relatively late, at about the time the blue-winged teals also appear, and males are soon actively engaged in aquatic head-pumping displays and noisy display flights. North American breeding grounds surveys in 2009 indicated a total population of 4.38 million birds, 92 percent above the long-term average (U.S.F.W.S., 2009a). The average annual hunter-kill estimate in the U.S. during the five years 2004–8 has been about 613,000 birds, and apparently has been slowly increasing since the 1960's, but the estimates were quite variable from year to year. Estimated total annual Canadian kills from 1990–1998 ranged from about 10,000–27,000. National Breeding Bird Surveys between 1966 and 2009 indicate that the species had a statistically significant population increase (1.0 percent annually) during that period.

Suggested Reading: Poston 1969; Johnsgard 1975; B.O.N.A. 217; Kear 2005.

Northern Pintail, *Anas acuta*

An abundant spring and fall migrant and a common summer resident throughout the three-state region, breeding locally in suitable habitats. Frequently over-winters in considerable numbers where open water occurs.

Migration: Sixty initial spring sightings in Nebraska range from January 18 to May 29, with a median of March 12. Half of the records fall within the period February 27 to March 20. Fifty-seven final fall sightings range from September 16 to December 31, with a median of November 19. Half of the records fall within the period November 6 to December 18. A state-level analysis of four decade-long periods of Christmas Bird Counts (1967-68 to 2006–7) extending from North Dakota to the Texas panhandle indicated a late-December population peak in northwestern Texas (Johnsgard and Shane, 2009).

Habitats: While on migration nearly all wetland habitats are used, ranging from flooded fields to large lakes and reservoirs. Breeding is also near wetlands ranging from small ponds to permanent marshes, but usually where the surrounding land is quite open and well drained.

Breeding Status: Breeds over most of the region, becoming more local and rarer southward. There were 30 confirmed nestings during breeding bird atlas surveys in South Dakota (Peterson, 1995), 21 in Nebraska (Mollhoff, 2001), and none in Kansas (Busby & Zimmerman, 2001).

Breeding Habitats: Over their vast range, northern pintails are associated with water types ranging from fresh to brackish, and from small temporary ponds to permanent marshes, but they are most abundant where there are open terrain surrounding areas of shallow water. In the Great Plains, stock ponds and similar wetlands with little or no vegetative cover are used more by pintails than by most other waterfowl.

Nest Location: Nests are invariably in dry, upland locations, often in dead plant growth of the previous year, at times with very little concealment. Many nests are placed in cover less than a foot high with no concealment on at least one side. Nests are often in shallow natural depressions, rendering them susceptible to flooding by heavy rains. They are lined with dark gray to brownish down.

Clutch Size and Incubation Period: From 5–11 eggs (68 nests in North Dakota averaged 7.9). The eggs are white to greenish yellow or grayish, with a smooth shell. The incubation period is 21 days, starting with the last egg. Single-brooded, but renests regularly if the first clutch is lost.

Time of Breeding: North Dakota egg dates range from April 13 to July 6, and dependent young have been reported from May 16 to September 17. Nests in South Dakota have been reported from April 18 to June 24 (Peterson, 1995). In Nebraska, young have been seen as early as May 28 (Mollhoff, 2001). Kansas egg dates are from April 21 to June 10, with a peak of egg-laying in early April.

Breeding Biology: Northern pintails form monogamous pair bonds during a prolonged period of social courtship, which continues as the birds migrate north in spring. Most or all females are paired by the time the birds arrive on their nesting grounds, and the pairs tend to become well spaced as they establish large home ranges. Females begin nesting very early, shortly after hillsides are free of snow, and like most ducks they complete their clutches at the rate of one egg per day. The incubation begins with the laying of the last egg, and by that time or shortly afterward the pair bond is broken. When the brood hatches, the female leads them to water, sometimes shifting ponds and moving them nearly a mile from where they were hatched. The fledging period is 47–57 days in South Dakota, and averages 41 and 46 days for females and males respectively in Manitoba.

Comments: Pintails are among the commonest breeding ducks in the region, along with mallards and blue-winged teal. North American breeding grounds surveys of pintails in 2009 indicated a total population of 3.22 million birds, 20 percent below the long-term average (U.S.F.W.S., 2009a). The average annual hunter-kill estimate in the U.S. during the five years 2004–8 has been about 442,000, but estimates have declined greatly from an annual high of nearly two million in the 1970's. Estimated total annual Canadian kills from 1990–1998 ranged from about 33,000–72,000. National Breeding Bird Surveys between 1966 and 2009 indicate that the species underwent a statistically significant population decline (2.4 percent annually) during that period.

Suggested Reading: Johnsgard 1975; Sowls 1978; B.O.N.A. 163; Kear 2005.

Green-winged Teal, *Anas crecca*

An abundant spring and fall migrant throughout the region. Breeds uncommonly in South Dakota, mostly in the northeast. Nebraska breeding is essentially limited to the northern part of the state and is concentrated in the Sandhills. Breeding in Kansas is rare.

Migration: Fifty-eight initial spring sightings in Nebraska range from January 1 to June 4, with a median of March 20. Half of the records fall within the period March 12–30. Fifty-five final spring sightings in Nebraska are from April 4 to June 10, with a median of May 10. Forty-six initial fall sightings are from August 3 to October 18, with a median of September 12. Forty-nine final fall sightings are from September 20 to December 31, with a median of November 2. A state-level analysis of four decade-long periods of Christmas Bird Counts (1967-68 to 2006–7) extending from North Dakota to the Texas panhandle indicated a late-December population peak in northwestern Texas (Johnsgard and Shane, 2009).

Habitats: Migrants are associated with almost all standing or slowly flowing aquatic habitats, and breeding normally occurs where a mixture of grassland, sedge meadows, and well-drained areas supporting shrubby or tall woody vegetation surrounds ponds or sloughs.

Breeding Status: Breeds over eastern South Dakota and the northern half of Nebraska. Kansas breeding records are few, and include one from the Salt Creek wetlands, Sumner County (Busby and Zimmerman, 2001), from Cheyenne Bottoms Wildlife Area in 1968, and one from Meade County (Thompson *et al.*, 2011). There were four confirmed nestings during breeding bird atlas surveys in South Dakota (Peterson, 1995), six in Nebraska (Mollhoff, 2001), and one in Kansas (Busby & Zimmerman, 2001).

Breeding Habitats: Green-winged teal breed in greatest numbers where there is a mixture of grassland, sedge meadows, and dry hillsides with low trees, brushy thickets, or open woods adjacent to ponds or sloughs. Grasslands lacking shrubs or thickets are not used as extensively as those with some woody cover, but the breeding ponds are often shallow and transient.

Nest Location: Nests are on dry land, usually well away from water and extremely well shaded by rushes, dense grasses, or shrubs. Low shrubs are apparently the preferred nesting site, especially those that offer excellent overhead concealment. The nest is a shallow excavation, lined with a very dark brown down.

Clutch Size and Incubation Period: Usually from 6–12 eggs (25 North Dakota nests averaged 8.6), varying from dull white to olive buff. The incubation period is 21–23 days. Single-brooded, but at least some renesting is known to occur.

Time of Breeding: Egg dates in North Dakota range from May 7 to July 28, and dependent young have been seen from June 20 to September 1. Nests in South Dakota have been reported from May 4 to June 5 (Peterson, 1995).

Breeding Biology: Green-winged teal are highly social and display over a long period of late winter and spring while forming their pair bonds, which are renewed annually. Pair-forming displays are numerous and elaborate and are highly animated. On reaching their breeding grounds, pairs spread out and establish home ranges that center on small ponds. Females select nest sites while accompanied by their mates, which usually remain attached to them until incubation is under way. After the clutch has hatched the female leads her young to shallow ponds, and they grow very rapidly. They fledge in no more than 44 days. Some Alaska fledging estimates are of as little as 35 days, but fledging is unusually rapid at such high latitudes, where long summer daylight allows for continuous feeding.

Comments: In spite of its small size, the green-winged teal is a very early spring migrant, appearing soon after mallards and northern pintails make their appearance. It is also a fairly late fall migrant, remaining long after the blue-winged teals have departed. North American breeding grounds surveys of green-winged teal in 2009 indicated a total population of 3.44 million birds, 79 percent above the long-term average (U.S.F.W.S., 2009a). Total U.S. kills have averaged about 1.72 million birds, and have exhibited a gradually increasing long-term trend-line since the 1960's. Estimated total annual Canadian kills from 1990–1998 ranged from about 93,000–145,000. National Breeding Bird Surveys between 1966 and 2009 indicate that the species underwent a statistically non-significant population decline (0.3 percent annually) during that period.

Suggested Reading: McKinney 1965; Johnsgard 1975; B.O.N.A. 193; Kear 2005.

Canvasback, *Aythya valisineria*

An uncommon to locally common spring and fall migrant throughout the region, and a local summer resident in northeastern South Dakota and the Nebraska Sandhills; with rare breeding in central Kansas.

Migration: Sixty-eight initial spring sightings in Nebraska are from February 12 to May 21, with a median of March 18. Half of the records fall within the period March 7 to March 30. Thirty-nine final fall sightings are from October 12 to December 31, with a median of November 14. Half of the records fall within the period October 29 through November 23. A state-level analysis of four decade-long periods of Christmas Bird Counts (1967-68 to 2006–7) extending from North Dakota to the Texas panhandle indicated a late-December population peak in northwestern Texas (Johnsgard and Shane, 2009).

Habitats: On migration this species uses marshes, rivers and shallow lakes rich in submerged weeds and similar vegetation. Prairie marshes with abundant emergent vegetation and some areas of open water are favored for nesting.

Breeding Status: Breeding is mostly confined to the glaciated areas of eastern South Dakota. There is also local breeding in the Nebraska Sandhills (especially Valentine and Crescent Lake national wildlife refuges). But there are only a few breeding records from Kansas, at Cheyenne Bottoms Wildlife Area, 1962–1973, and Quivira N.W.R., 1980 (Thompson *et al.*, 2011). There were ten confirmed nestings during breeding bird atlas surveys in South Dakota (Peterson, 1995), none in Nebraska (Mollhoff, 2001), and none in Kansas (Busby & Zimmerman, 2001).

Breeding Habitats: Canvasbacks are most abundant on shallow prairie marshes that are surrounded by cattails, bulrushes, and similar vegetation, with open water for landing and taking off and little or no wooded area around the shoreline.

Nest Location: Nests are constructed over water, among emergent vegetation that is 1–4 feet tall and composed of bulrushes (usually hardstem bulrush) or cattails, with phragmites used less often. Nests are usually 10–15 yards from areas of open water that are at least 50 feet square. The nest bowl is usually well lined with pearly gray down having inconspicuous white tips.

Clutch Size and Incubation Period: From 6–16 eggs (26 North Dakota nests averaged 9.9), with larger clutches of "dump nests" not infrequent. The eggs are grayish olive to greenish. The incubation period is about 24 days. Single-brooded, but renesting occurs frequently.

Time of Breeding: North Dakota egg dates range from April 28 to July 15, and dates of dependent young are from May 29 to September. Nests in South Dakota have been reported from May 5 to July 19 (Peterson, 1995). In Nebraska, young have been seen as early as June 6 (Mollhoff, 2001).

Breeding Biology: Canvasbacks renew their pair bonds annually, and courtship is usually intense as the birds are returning to their nesting grounds. Several aquatic displays, including cooing calls and head-throw displays, are conspicuous then. As pairs form, they separate from the flocks and seek out nesting areas in smaller and shallower ponds than those used for courting. In densely populated areas a substantial amount of nest parasitism occurs among canvasbacks and between canvasbacks and redheads. Although parasitic redheads are prone to lay their eggs in canvasback nests, the latter usually lay eggs only in the nests of other canvasbacks. Thus mixed-species broods sometimes occur, but parasitized nests are less successful than non-parasitized ones. The fledging period is 8–9 weeks.

Comments: Canvasbacks have traditionally been regarded as regal ducks; their long sloping bills and robust outlines set them apart from other ducks. They are larger than redheads, generally paler in both sexes, and lack the high rounded head profile of that species. North American breeding grounds surveys in 2009 indicated a total population of about 700,000 birds, or six percent above the long-term average (U.S.F.W.S., 2009a). The average annual hunter-kill estimate in the U.S. during the five years 2004–8 has been about 68,000 birds, but both the yearly figures and long-term trends since the 1960's have been highly variable, perhaps reflecting varying degrees of protection from hunters. Estimated total annual Canadian kills from 1990–1998 ranged from about 5,000–13,000. National Breeding Bird Surveys between 1966 and 2009 indicate that the species underwent a statistically non-significant population decline (0.5 percent annually) during that period.

Suggested Reading: Hochbaum 1944; Johnsgard 1975; B.O.N.A. 659; Kear 2005.

Redhead, *Aythya americana*

A common spring and fall migrant throughout the region. It is a lo-cally common summer resident in the Nebraska Sandhills west to Garden County, as well as breeding in the Rainwater Basin of Nebraska. Breeding is regular in South Dakota and is at least occasional in central Kansas.

Migration: Sixty initial spring sightings in Nebraska range from Feb-ruary 9 to May 25, with a median of March 13. Half of the sightings fall within the period March 1 through March 20. Fifty-six final fall sightings are from October 9 to December 1, with a median of November 9. Half of the records fall within the period October 28 to November 19. A state-level analysis of four decade-long periods of Christmas Bird Counts (1967-68 to 2006–7) extending from North Dakota to the Texas panhandle indicated a late-December population peak in northwestern Texas (Johnsgard and Shane, 2009).

Habitats: Migrants are found on large prairie marshes, lakes and res-ervoirs, especially where submerged vegetation is abundant. Nesting typi-cally occurs on marshes at least an acre in size, having both open areas and stands of emergent vegetation.

Breeding Status: Breeds in South Dakota, mainly east and north of the Missouri River, as well as in the Sandhills of Nebraska. There are a few nest-ing records from Kansas, from Quivira N.W.R., Cheyenne Bottoms Wildlife Area, the Slate Creek wetlands of Sumner County, and from Pratt and Sedg-wick counties Thompson et al, 2011). There were seven confirmed nestings during breeding bird atlas surveys in South Dakota (Peterson, 1995), seven in Nebraska (Mollhoff, 2001), and three in Kansas (Busby & Zimmerman, 2001).

Breeding Habitats: Redheads are similar to canvasbacks in their habitat needs, but occur on more alkaline marshes. They usually nest in marshes at least an acre in area, with about 10-25 percent of the surface open water and emergent vegetation 20–40 inches tall.

Nest Location: Nests are built in emergent vegetation, in water about a foot deep and in vegetation 20-40 inches tall. They are placed within 50 yards of open water, often less than five yards away. Cattails and hard stem bulrushes are favored nesting sites. The nest bowl is lined with down that varies from white to medium gray .

Clutch Size and Incubation Period: From 8–15 eggs (74 North Dakota nests averaged 10.2), but parasitic and "dump nesting" often makes clutches abnormally large. The eggs are usually creamy white, rarely greenish to buffy. The incubation period is 24-28 days, starting with the last egg.

Time of Breeding: North Dakota egg dates range from May 5 to Au-gust 10, and dates of dependent young range from June 14 to October 17.

Nests in South Dakota have been reported from May 26 to June 5 (Peterson, 1995). In Nebraska, young have been seen as early as June 4 (Mollhoff, 2001).

Breeding Biology: Redheads have seasonal pair bonds, established each winter and spring. Their displays and associated behavior are much like those of canvasbacks, and the two species often associate. On reaching their nesting grounds, pairs establish home ranges that typically include nest-site potholes and waiting-site potholes, often shared with other pairs. Nest parasitism among redheads is high in most areas, and they drop eggs in the nests of a large variety of other marsh birds, although not all females are parasitic nesters. Males abandon their mates early in incubation and often fly elsewhere to molt. In Iowa the young fledge at 70-84 days of age, and shorter fledging periods have been reported for Canada. In Iowa there is also a moderate amount of renesting, but little or none occurs in Canada.

Comments: Redheads are considerably more common than canvasbacks in the region, and can usually by seen during summer at Crescent Lake and Valentine refuges. Neck-stretching and head-throw displays are common in spring, accompanied by soft cat-like meowing calls. North American breeding grounds surveys in 2009 indicated a total population of 1.04 million birds, or 62 percent above the long-term average (U.S.F.W.S., 2009a). The average annual hunter-kill estimate in the U.S. during the five years 2004–8 has been about 148,000 birds, but the yearly estimates have been fairly variable since the 1960's, perhaps reflecting varying degrees of protection from hunters. Estimated total annual Canadian kills from 1990–1998 ranged from about 11,000–22,000. National Breeding Bird Surveys between 1966 and 2009 indicate that the species underwent a statistically non-significant population increase (0.7 percent annually) during that period.

Suggested Reading: Low 1945; Johnsgard 1975; B.O.N.A. 695; Kear 2005.

Ring-necked Duck, *Aythya collaris*

An uncommon to common spring and fall migrant throughout the region. becoming less common to the west. Nebraska is outside the current breeding range of this species.

Migration: Forty-two initial spring sightings in Nebraska are from February 12 to May 25, with a median of March 21. Half of the records fall within the period March 7 to March 30. Twenty-six final spring sightings in Nebraska are from March 24 to May 30, with a median of April 21. Twenty-seven initial fall sightings are from September 17 to December 7, with a median of October 12. Twenty-three final fall sightings are from October 27 to December 31, with a median of November 17. A state-level analysis of four decade-long periods of Christmas Bird Counts (1967-68 to 2006–7) extending from North Dakota to the Texas panhandle indicated a late-December population peak in northwestern Texas (Johnsgard and Shane, 2009).

Habitats: Migrants are found on large prairie marshes, lakes and reservoirs, but prairie marshes are only secondary breeding habitats. Rather acidic swamps and bogs, surrounded by shrubby covers are the primary breeding habitat.

Breeding Status: There are scattered breeding records for South Dakota, with at least three confirmed records (Marshall and Codington counties) during breeding bird atlas surveys (Peterson, 1995). Other than these confirmed nestings during breeding bird atlas surveys in South Dakota, there were none in Nebraska (Mollhoff, 2001) or Kansas (Busby & Zimmerman, 2001). Nesting in that state occurs in several northeastern counties (Tallman, Swanson and Palmer, 2002). It apparently bred historically in the Nebraska Sandhills, as it was reported by Oberholser and McAtee (1920) to breed in Brown, Cherry, Garden and Morrill counties.

Breeding Habitats: The primary habitat of ring-necked ducks is sedge-meadow marshes and bogs, with freshwater marshes used secondarily. Acidic or freshwater wetlands are preferred to brackish ones, and the birds especially frequent ponds that support water lilies and are surrounded by shrubby cover.

Nest Location: Nests are very frequently on floating islands in bogs or on hummocks of vegetation in open marshes. Rarely, nests are on dry land well away from water, but most are close to water of swimming depth and within 15 yards of water open enough for landing and taking off. Both sedges and brushy cover are used for specific nest cover, and small clumps are apparently used more often than larger ones. Nests are lined with sooty brown down having white centers.

Clutch Size and Incubation Period: From 6–14 eggs, with an average of about 10. The eggs are olive buff and smooth. The incubation period ranges from 25 to 29 days, averaging about 27 days after the laying of the last egg. Single-brooded, but a persistent renester.

Time of Breeding: North Dakota egg dates range from May 30 to July 20, and dates of dependent young are from June 29 to September 7. Nests in South Dakota have been reported from June 23 to July 26 (Peterson, 1995).

Breeding Biology: Pair bonds in ring-necked ducks start to become established on the wintering grounds through social display that begins in January and February, but some displays persist until the birds arrive on their nesting grounds. Display patterns are much like those of redheads and canvasbacks, in spite of plumage differences. Pairs become spaced out over the breeding grounds but show little aggression when they come into contact, and nests are often close together on islands. The pair bond is usually broken near the end of the incubation period, and females raise their broods alone, on ponds often largely covered by water lilies. By the end of the fledging period of 7–8 weeks the female has begun her flightless period and family bonds terminate.

Comments: This species, often called the "ring-billed duck," is closely related to the redhead, as is obvious from its downy plumage. However, its breeding habitats are quite different from those of redheads and canvasbacks. Eastern North American breeding grounds surveys in 2009 indicated a total population of 551,000 birds (U.S.F.W.S., 2009a). The average annual hunter kills in the U.S. during the five years 2004–8 have been about 513,000 birds, and have exhibited a long-term progressive increase since the 1960's. Estimated total annual Canadian kills from 1990–1998 ranged from about 57,000–110,000. National Breeding Bird Surveys between 1966 and 2009 indicate that the species underwent a statistically non-significant population increase (1.5 percent annually) during that period.

Suggested Reading: Mendall 1958; Johnsgard 1975; B.O.N.A. 329; Kear 2005.

Lesser Scaup, *Aythya affinis*

A common to abundant spring and fall migrant throughout the region. An uncommon local breeder in northeastern South Dakota. and an occasional summer resident in the Nebraska Sandhills. Over-winters locally where open water is present.

Migration: Sixty-nine initial spring sightings in Nebraska are from February 12 to May 20, with a median of March 19. Half of the records fall within the period March 5–25. Forty-three final spring Nebraska records are from March 10 to June 6, with a median of May 11. Forty-five initial fall sightings are from July 20 to December 15, with a median of October 18. Thirty-one final fall sightings are from November 22 to December 31, with a median of December 14. A state-level analysis of four decade-long periods of Christmas Bird Counts (1967-68 to 2006–7) extending from North Dakota to the Texas panhandle indicated a late-December population peak in northwestern Texas (Johnsgard and Shane, 2009).

Habitats: Deeper marshes, reservoirs, borrow-pits and lakes are commonly used by migrating birds. Prairie marshes surrounded by partially wooded uplands are favored for breeding, especially those supporting large populations of amphipods (scuds).

Breeding Status: There were at least three confirmed records during breeding bird atlas surveys in South Dakota (Peterson, 1995), but none in Nebraska (Mollhoff, 2001) or Kansas (Busby & Zimmerman, 2001). Nesting in South Dakota occurs in several northeastern counties, including Day, Edmunds and McPherson (Tallman, Swanson and Palmer, 2002). It has occasionally bred in the Nebraska Sandhills (probably Brown, Cherry, Garden and Morrill counties), and there is a Kansas breeding record from Cowley County in 1928 (Thompson *et al.*, 2011).

Breeding Habitats: Prairie marshes, glacial potholes, and ponds or lakes in partially wooded parklands are the major habitat of lesser scaups. Favored ponds are usually slightly to moderately brackish and vary from semi permanent to permanent. Those supporting high populations of amphipods and aquatic insects are most heavily used.

Nest Location: Nests are on dry land but usually fairly close to shore. Islands on lakes, especially those with grassy or weedy cover, are especially preferred. Nests are usually within 50 yards of water, and the cover is 1–2 feet tall. In some boggy areas the nests are placed on floating sedge mats. The nest cup is lined with very dark brown down with inconspicuous white centers.

Clutch Size and Incubation Period: From 7–12 eggs (25 North Dakota nests averaged 9.4). The eggs are greenish to olive buff with a smooth but not glossy surface. The incubation period is 21–26 days, with an average of 24.8 days from the date of the last egg. Single-brooded, but a regular renester (up to three renesting attempts have been reported).

Time of Breeding: North Dakota egg dates range from May 21 to August 10, with more than 80 percent of the records in June. Nests in South Dakota have been reported from June 12–25 (Peterson, 1995).

Breeding Biology: Lesser scaups form pair bonds that persist for a single season, during a prolonged period of winter and spring social display. Pairs establish relatively large but poorly defined home ranges, often centering on marshes 2–5 acres in area that include some deep water. Females build their nests alone and are abandoned by their mates shortly after they begin incubation. Although scaups often nest in or near gull colonies, the gulls sometimes prey severely on ducklings, and much brood disruption is typical. Females often desert their ducklings early to molt, and large broods consisting of ducklings from several families are frequent in some areas. The fledging period is about 47–50 days, a relatively short period for diving ducks.

Comments: This "bluebill" is a very common spring and fall migrant, and is perhaps the most abundant of the diving ducks in the region. Males have bright blue bills in spring; like male ruddy ducks this blue coloration is probably caused by light refraction effects, rather than by blue pigmentation. Breeding grounds surveys in 2009 indicated a total population of 4.2 million scaups of both species, or 18 percent below the long-term average (U.S.F.W.S., 2009a). Nearly 90 percent of the scaups surveyed nationally are probably lesser scaups (Bellrose, 1980). The average annual hunter-kill estimate in the U.S. during the five years 2004–8 has been about 235,000 birds. However, a continent-wide population decline has been occurring since the 1980's, and average kill estimates have exhibited a long-term decline from a peak of about 600,000 during the 1980s. Estimated total annual Canadian kills from 1990–1998 ranged from about 41,000–71,000. National Breeding Bird Surveys between 1966 and 2009 indicate that the species underwent a statistically significant population decline (1.8 percent annually) during that period.

Suggested Reading: Trauger 1971; Johnsgard 1975; Hines 1977; B.O.N.A. 338; Kear 2005.

Bufflehead, *Bucephala albeola*

A common to uncommon migrant throughout the region. Occasionally stragglers remain through the summer. Breeding occurs locally in South Dakota, and migrants occur throughout the Great Plains.).

Migration: Fifty-three initial spring sightings in Nebraska are from February 21 to May 1, with a median of March 18. Half of the records fall within the period March 6–24. Thirty-eight final spring sightings in Nebraska are from March 15 to May 29, with a median of April 21. Thirty-four initial fall sightings are from August 14 to December 16, with a median of October 19. Thirty-one final fall sightings are from October 29 to December 31, with a median of November 24. A state-level analysis of four decade-long periods of Christmas Bird Counts (1967-68 to 2006–7) extending from North Dakota to the Texas panhandle indicated a late-December population peak in northwestern Texas (Johnsgard and Shane, 2009).

Habitats: Lakes, reservoirs and deeper marshes are used by migrating birds.

Breeding Status: There were two confirmed nesting during breeding bird atlas surveys in South Dakota (Peterson, 1995), but none in Nebraska or Kansas. The only confirmed nesting records from South Dakota are from Brookings and Roberts counties, but six other counties have reported breedings. (Tallman, Swanson and Palmer, 2002. There are no fully accepted breeding records for Kansas. but a brood of four reported in Pratt County in 1988 would seem convincing (Thompson *et al*, 2011).

Habitats: Buffleheads are associated with ponds or lakes in or near open woodland, where nesting sites are available and where there is an abundance of aquatic invertebrates. Moderately deep and fertile lakes, with open shore-lines and sparse reed beds that have nest sites nearby, are especially favored.

Nest Location: Nests are almost always in cavities made by woodpeck-ers, primarily flickers. Cavities that are 3–20 feet above the ground or water level are used, especially those in dead trees that are either standing in wa-ter or are very close to it. The birds will also nest in artificial cavities, with entrances about three inches wide and internal diameters of about seven inches. The nest is lined with a down that is pale grayish to brownish with indistinct lighter centers.

Clutch Size and Incubation Period: From 5–16 eggs (averaging about nine in initial clutches and seven in late or renesting efforts). The eggs are creamy to olive buff with a slightly glossy surface. The incubation period ranges from 29–31 days, usually 30 days. Single-brooded; renesting is ap-parently uncommon and possibly does not occur.

Time of Breeding: North Dakota egg dates are from June 6–13, and de-pendent young have been seen from June 15 to August 15. A nest in South Dakota was reported with eggs from May 22 to June 16 (Peterson, 1995).

Breeding Biology: Buffleheads form seasonal pair bonds during a pro-longed period of courtship that extends from winter through the spring Mi-gration: It is not known how often males re-mate with mates of the previ-ous year, but females have a strong tendency to return to the place where they nested previously, and they often nest in the same cavity. Competition for nest sites among buffleheads and with other hole-nesting birds such as starlings and tree swallows makes nest-site availability an important facet of their biology. Males abandon their mates and often leave the area shortly af-ter incubation gets under way, and the females leave their nests occasion-ally during incubation to feed. The young remain in the nest 24–36 hours after hatching, then at their mother's signal they jump down to the ground and leave as a group, usually during the morning. The fledging period is about 50–55 days, during most of which the female keeps the young within a brood territory. However, some brood transfers and formations of multi-ple broods have been reported.

Comments: Easily one of the most beautiful of America's ducks, these birds tend to appear on the same kinds of wetlands as goldeneyes and mer-gansers, the males sparkling like gigantic snowflakes on the water, and the tiny, drab females almost invisible by contrast. Nesting occurs in the cav-ities made by flickers and other small woodpeckers. There is a population estimate of one million birds for all of North America (Wetlands Interna-tional, 2002). The average annual hunter-kill estimate in the U.S. during the five years 2004–8 has been about 189,000 birds, and kill estimates have been gradually increasing since the 1960's. Estimated total annual Canadian kills from 1990–1998 ranged from about 18,000–37,000.

Suggested Reading: Erskine 1972; Johnsgard 1975; B.O.N.A. 67; Kear 2005.

Hooded Merganser, *Lophodytes cucullatus*

An uncommon to occasional spring and fall migrant in eastern parts of the region, and an occasional to rare migrant in western parts. In all three states it is a rare and local breeder.

Migration: Seventy-four initial spring sightings in Nebraska range from January 16 to May 30, with a median of March 26. Half of the records fall within the period March 1328. Fourteen final spring Nebraska records are from March 19 to May 30, with a median of April 25. Sixteen initial fall sightings are from September 14 to November 27, with a median of November 5. Nineteen final fall sightings are from November 6 to December 17, with a median of November 22. A state-level analysis of four decade-long periods of Christmas Bird Counts (1967-68 to 2006–7) extending from North Dakota to the Texas panhandle indicated a late-December population peak in Oklahoma (Johnsgard and Shane, 2009).

Habitats: Migrants are found on clear-water rivers, lakes, reservoirs and deeper marshes. Breeding is usually on rivers, creeks and oxbows bordered by woods and supporting good populations of fish.

Breeding Status: There are apparently valid early records of Nebraska nesting from Lancaster, Gage, and Cuming counties (Bruner, Wolcott and Swenk, 1904), and the author observed a female and five newly fledged young at Twin Lakes, Seward County, in July, 1995. Two other breeding re-cords for eastern Nebraska also exist (*Nebraska Bird Review* 73:50). There was at least one confirmed nesting during breeding bird atlas surveys in Kansas (Busby & Zimmerman, 2001), and at least three for South Dakota (Peterson, 1995). Nesting in South Dakota has occurred in at least three eastern counties (Tallman, Swanson and Palmer, 2002), and there are also breeding records for at least six eastern counties (Thompson *et al.*, 2011). There were three confirmed nestings during breeding bird atlas surveys in South Dakota (Peterson, 1995), none in Nebraska (Mollhoff, 2001) and one in Kansas (Busby & Zimmerman, 2001).

Breeding Habitats: In North Dakota, hooded mergansers are associated with rivers, creeks, and oxbows bordered by woods and supporting large populations of small fish. In general, they prefer clear streams that have woods nearby to provide nesting cavities, especially streams that have sandy or cobble Bottoms rather than mud Bottoms and are not very deep or murky enough to interfere with underwater vision.

Nest Location: Cavities or nesting boxes close to water are used more often than those well away from it. The cavity entrance may be near the ground, or up to 60 feet high. Hooded mergansers readily accept nesting boxes that have been set out for wood ducks, so their general requirements for cavity size are probably very similar. Both species often use the same boxes where they occur together, resulting in mixed clutches that may be incubated by either species. The down lining of hooded merganser nests is pale gray with white centers (as in wood ducks), but the associated breast feathers are narrow and off-white rather than wide and white.

Clutch Size and Incubation Period: From 7–13 eggs (five North Dakota clutches averaged 8.7). The eggs are white with a glossy surface and are slightly larger (over 40 mm wide) and more rounded than wood duck eggs. The incubation period is about 33 days. Single-brooded, with no evidence of renesting tendencies.

Time of Breeding: North Dakota egg dates are for June and early July, with young (or hatched eggs) reported from June 18 to August 7. Nests in South Dakota have been reported from May 28 to June 6 (Peterson, 1995).

Breeding Biology: Hooded mergansers first form pairs in their second winter of life and thereafter establish pair bonds annually during a prolonged courtship period. On return to their nesting areas, the females usually find nest sites near their former nest areas and many nest in the very same locations. These locations are often within a few miles of where the females hatched. Pair bonds are probably disrupted early in the incubation period, and incubating females leave their nests two or three times a day to forage. The newly hatched young are taken to rearing areas, usually rivers but sometimes standing-water habitats such as beaver ponds. They fledge in about 70 days.

Comments: This species seems to be increasing, as it was rarely reported during the 1960's and 1970's, but now it is a regular spring migrant in eastern areas of Nebraska. Perhaps it has benefited from nest-box erection programs for wood ducks, as it often will choose such places for laying its eggs. Range-wide population estimates are not available, but hunter harvest data suggest a population during the 1990's of at least 270,000–385,000. The average annual hunter-kill estimate in the U.S. during the five years 2004–8 has been about 84,800 birds, and estimates have exhibited an increasing trend-line since the 1960's. Estimated total annual Canadian kills from 1990–1998 ranged from about 14,000–29,000. National Breeding Bird Surveys between 1966 and 2009 indicate that the species underwent a statistically significant population increase (3.6 percent annually) during that period.

Suggested Reading: Morse, Jakabosky & McCrow 1969; Johnsgard 1975; B.O.N.A. 98; Kear 2005.

Common Merganser, *Mergus merganser*

A regular spring and fall migrant throughout the region, varying in abundance from very common to occasional. Over-winters commonly where open water persists; stragglers sometimes remain through the summer.

Migration: Fifty initial spring sightings in Nebraska are from January 14 to April 25, with a median of March 9. Half of the records fall within the period March 3–27. Thirty-nine final spring sightings in Nebraska are from March 4 to May 30, with a median of April 6. Thirty-eight initial fall sightings are from September 18 to December 31, with a median of November 13. Thirty-six final fall sightings are from November 20 to December 31, with a median of December 17. Larger numbers stage during early spring at Harlan County Reservoir; one estimate was of 200,000 birds, and similar numbers have been estimated for Glen Elder Reservoir, Kansas (*Nebraska Bird Review 73:67*). A state-level analysis of four decade-long periods of Christmas Bird Counts (1967-68 to 2006–7) extending from North Dakota to the Texas panhandle indicated a late-December population peak in Kansas (Johnsgard and Shane, 2009).

Habitats: Migrants and wintering birds are found on rivers, lakes, reservoirs, and any other large wetlands supporting fish populations. Most nesting occurs on forest-lined lakes and ponds near rivers, but rarely nesting occurs in treeless areas in rock crevices or other natural cavities.

Breeding Status: Evidence of breeding in Nebraska includes a brood sighted in Custer County in 1968 (*Nebraska Bird Review* 37:45) and one reported from the North Platte Valley west of Lake McConaughy (*Nebraska Bird Review* 62:105). Summering birds are regular in the Lake Alice area of Scotts Bluff County, and nesting has been suspected. There was at least one confirmed breeding record during breeding bird atlas surveys in South Dakota (Peterson, 1995), but none during such surveys in Nebraska (Mollhoff, 2001) or Kansas (Busby & Zimmerman, 2001). Nesting in South Dakota has occurred in Pennington and Charles Mix counties (Tallman, Swanson and Palmer, 2002).

Breeding Habitats: Common mergansers are mostly associated with forest-lined lakes and ponds near rivers that support high fish populations and are clear enough to allow for visual foraging.

Nest Location: In our region nesting is mostly confined to cavities in hardwood trees, but in areas without trees large enough to provide nest cavities the birds will nest under boulders, in buildings, or under dense brush. Concealment and darkness in the nest cavity are prime requirements. The birds will also nest in artificial cavities such as nest boxes with entrances 4¼ inches in diameter, an internal cavity about ten inches wide, and a depth of about 20 inches below the entrance. Nests are usually lined with some down, which is nearly white, unlike the dark down of red-breasted mergansers.

Clutch Size and Incubation Period: From 7–16 eggs, often 9–10. The eggs are pale cream to buffy and lack a glossy surface. The incubation period is 32–35 days. Single-brooded.

Time of Breeding: Eggs in North Dakota nests have been reported from May 26 to June 25. Broods in South Dakota have been reported in late May (Peterson, 1995).

Breeding Biology: During fall and winter, these mergansers usually stay in small flocks that sometimes feed cooperatively, but as spring approaches much time is spent in social display and in establishing pair bonds, and flock sizes decrease. Females remain fairly gregarious while looking for nest sites and often nest close together. Probably some dump nesting occurs in locations where nest cavities are limited. The males usually leave their mates before hatching but on rare occasions have been seen with broods. The young are led to water a day or two after hatching, and the brood is usually raised in shallow rivers. At times the female carries part of her brood on her back, especially when they are frightened. The fledging period is 60–70 days.

Comments: It is likely that more nesting records of this splendid species will accrue as time passes, since summering birds are increasingly common in Nebraska. The birds need clear waters with a good fish population; hunters and fishermen tend to hate this and other merganser" species because of their fishy taste and appetites. However, the birds generally eat slow-swimming prey rather than trout and other game fish. World population estimates of this widely distributed northern hemisphere (Holarctic) species include 640,000 birds in North America (Kear, 2005). The average annual hunter-kill estimate in the U.S. during the five years 2004–8 has been about 18,600 birds, and estimates have been progressively increasing since the 1960's. Estimated total annual Canadian kills from 1990–1998 ranged from about 12,000–20,000. National Breeding Bird Surveys between 1966 and 2009 indicate that the species underwent a no statistically significant population decline (1.0 percent annually) during that period.

Suggested Reading: White 1957; Johnsgard 1975; B.O.N.A. 442; Kear 2005.

Ruddy Duck, *Oxyura jamaicensis*

A common spring and fall migrant throughout the region, and an uncommon and very local summer resident in the Nebraska Sandhills and Rainwater Basin, and in northeastern South Dakota.

Migration: Sixty-seven initial spring sightings in Nebraska are from February 12 to June 9, with a median of April 3. Half of the records fall within the period March 14 to April 19. Fifty-nine final fall Nebraska records are from August 30 to December 31, with a median of November 27. Half of the records fall within the period October 10 to November 27. A state-level analysis of four decade-long periods of Christmas Bird Counts (1967-68 to 2006–7) extending from North Dakota to the Texas panhandle indicated a late-December population peak in northwestern Texas (Johnsgard and Shane, 2009).

Habitats: Migrants may be found on lakes, reservoirs, larger marshes and similar habitats offering considerable open water and mud-bottom feeding areas. Breeding is on prairie marshes having stable water levels and an abundance of emergent vegetation, along with some areas of open water.

Breeding Status: Breeding occurs in South Dakota in the glaciated areas to the east and north of the Missouri River, and in the deeper marshes of the Nebraska Sandhills and Rainwater Basin. There were six confirmed nestings during breeding bird atlas surveys in South Dakota (Peterson, 1995), and two in Kansas (Busby & Zimmerman, 2001). Breeding in Kansas is very local, with definite nesting records from Barton, Stafford, and Grant counties (Thompson *et al.*, 2011). It is considered a common breeder at both Valentine N.W.R. and Crescent Lake N.W.R. in Nebraska. There were six confirmed nestings during breeding bird atlas surveys in South Dakota (Peterson, 1995), eight in Nebraska (Mollhoff, 2001) and two in Kansas (Busby & Zimmerman, 2001).

Breeding Habitats: Permanent or semi-permanent prairie marshes having stable water levels and an abundance of emergent vegetation, especially cattails and hardstem bulrushes, are prime ruddy duck habitat. The marshes must have some open water close to the nesting cover provided by the emergent vegetation or be connected with open water by muskrat channels.

Nest Location: Hardstem bulrush stands are optimum nesting cover for ruddy ducks, which build platforms or floating nests in water up to three feet deep. Dense cover is preferred to sparse cover, and bulrushes are preferred to cattails or other emergent plants. Bulrushes that can be readily bent over to form the nest are preferred; species with stiff, tough stalks are little used. Many nests have canopies and ramps for easy access. Usually there is a little down present as a lining, which is very light colored with white centers.

Clutch Size and Incubation Period: From 3–15 eggs (68 North Dakota nests averaged 7.7), but dump nests or parasitically laid eggs often cause inflated clutch sizes. The eggs are white with a chalky and granular surface. The incubation period is 23 days. Single-brooded, with limited evidence of renesting.

Time of Breeding: North Dakota egg dates range from May 29 to August 21, and dependent young have been seen from July 3 to October 14. Nests in South Dakota have been reported as early as May 24 (Peterson, 1995). In Nebraska, young have been seen as early as July 14 (Mollhoff, 2001). Kansas egg records are for the period May 30 to July 30, and broods have been seen from June 10 to August 18.

Breeding Biology: Ruddy ducks apparently mature in their first year, though not all females are thought to breed as yearlings. Pair bonds are rather weak, and much display is related to territorial advertisement rather than courtship itself. Females show little or no pair-forming or pair-maintaining behavior, although males may remain in the vicinity of their mates after they have begun nesting. Some even persist in remaining with them after the brood has hatched, though they do not "assist" in rearing the brood. The young are highly precocious and are soon independent, so that broods often become scattered before they fledge, about 6–7 weeks after hatching.

Comments: This wonderful little duck is nearly everybody's favorite; from its sky-blue bill to its long, spiky tail it advertises its uniqueness. From the time it finally arrives in spring and begins its bizarre displays through the summer brooding period it is hard to imagine a more interesting bird to watch. Like other stiff-tailed ducks it feeds almost entirely on small fly larvae that live in the muddy bottoms of ponds, using its sensitive bill to locate its prey. The North American population has been estimated at about 500,000 birds (Kear, 2005). The average annual hunter-kill estimate in the U.S. during the five years 2004–8 has been about 28,200 birds, and estimates have exhibited a long-term decline since the 1960's. Estimated total annual Canadian kills from 1990–1998 ranged from about 700–4,000. National Breeding Bird Surveys between 1966 and 2009 indicate that the species underwent a statistically non-significant population increase (0.5 percent annually) during that period.

Suggested Reading: Low 1941; Johnsgard 1975, 1996; B.O.N.A. 696; Kear 2005.

ORDER PODICPEDIFORMES - GREBES

Family Podicipedidae - Grebes

Pied-billed Grebe, *Podilymbus podiceps*

A common spring and fall migrant and local summer resident throughout the three-state region. It breeds almost throughout the Plains States.

Migration: A total of 116 initial spring sightings in Nebraska range from February 27 to June 10, with a median of April 5. Half of the sightings fall within the period March 24–April 22. Eighty-four final fall sightings are from August 21 to December 6, with a median of November 6. Half of the sightings fall within the period October 10–November 16. A state-level analysis of four decade-long periods of Christmas Bird Counts (1967-68 to 2006–7) extending from North Dakota to the Texas panhandle indicated a late-December population peak in Oklahoma (Johnsgard and Shane, 2009).

Habitats: Breeding occurs on small ponds, river impoundments and lakes, ranging from quite small to large, but always those having extensive stands of heavy emergent vegetation and adjacent areas of open water.

Breeding Status: Virtually pandemic throughout the region. It is a locally common breeder in South Dakota east of the Missouri River, in the Nebraska Sandhills, and in central and western Kansas (confirmed breedings in Barton, Kearny and Stafford counties; summer records from at least five other counties). There were 26 confirmed nestings during breeding bird atlas surveys in South Dakota (Peterson, 1995), 23 in Nebraska (Mollhoff, 2001) and 18 in Kansas (Busby & Zimmerman, 2001). Until the modification of Nebraska's Kingsley Dam for hydropower generation in 1982, breeding occurred on Lake Ogallala/Keystone, Keith County.

Breeding Habitats: In North Dakota, pied-billed grebes breed on seasonal or permanent ponds ranging from fresh to moderately brackish, on river impoundments and lakes, particularly those having extensive stands of emergent vegetation with adjacent areas of open water. Compared with eared and horned grebes, this species occupies a wider variety of pond types but is always associated with heavy emergent vegetation, and its distinctive vocalizations adapt it well to establishing territories in low-visibility habitats.

Nest Location: Nests float and are usually in semi-open to dense emergent vegetation, frequently bulrushes. Grasses (whitetop, mannagrass), sedges, cattails, and bur reeds are also sometimes used in North Dakota, and water depths at more than 80 sites averaged 25 inches, ranging from 11 to 37 inches. Two or more nest platforms may be constructed by a pair, from four to ten yards apart, often at the edge of vegetation to allow an underwater approach to the nest. However, in Iowa, 138 nests averaged about 26 feet from nest to open water.

Clutch Size and Incubation Period: Usually from 4–7 eggs, bluish white to greenish white initially, but gradually becoming stained with brown. A sample of 74 nests in North Dakota averaged 6.7 eggs per clutch; 97 successful nests in Iowa averaged 6.18 eggs. The incubation period is 23 days, probably starting with the first egg, since hatching is usually spread over several days. Most incubation is by the female, but both sexes participate.

Time of Breeding: Extreme egg dates in North Dakota range from May 7 to August 20. Nests in South Dakota have been reported from May 1 to July 22 (Peterson, 1995). In Nebraska, nests with eggs have been seen as early as June 15, and young as early as July 4 (Mollhoff, 2001). In Kansas, egg records are from May 1 to July 14, and flightless young seen from June 18 to August 7.

Breeding Biology: Pairs are very territorial; in Iowa the territory consisted of an arc of about 150 feet around the nest, with the male defending the area, but the pair shared wetlands outside territorial limits with other birds. The birds are highly vocal, and territorial displays are evidently as much acoustic as postural. Courtship consists of advertising calls, "ripple dives," and fairly simple mutual displays : a "circle" display in which the birds face one another and rotate, and a "pirouette ceremony," in which the pair meets in upright postures, rise high in the water, and pirouette as they utter mechanical notes. Copulation normally occurs on the nest platform, and during treading the passive bird brings its head back and strokes it against its partner's breast. Both parents care for the young, which regularly ride on the backs of their parents. The fledging period is 35–37 days (Fjeldsa, 2004). By three weeks of age the young are fairly independent, which may allow adults to begin a second clutch while the young are still flightless.

Comments: This grebe is the commonest species in the region, and one that rarely strays far from reedy or weedy shorelines. It often dives vertically when alarmed, with at most its head remaining above water. Its bill is adapted for eating crustaceans and other invertebrates rather than fish, and it is never seen in flocks like the other regional grebes. National Breeding Bird Surveys between 1966 and 2009 indicate that the species underwent a statistically significant population decline (0.3 percent annually) during that period.

Suggested Reading: Glover 1953; Palmer 1962, Johnsgard, 1987; B.O.N.A. 410; Fjeldsa, 2004.

Horned Grebe, *Podiceps auritus*

An uncommon spring and fall migrant throughout The three-state region, and an accidental summer resident. It has reportedly bred in Cherry County, Nebraska, but the closest area of regular breeding is in northern South Dakota.

Migration: Sixty-two initial spring sightings in Nebraska range from February to June 4, with a median of April 16. Twenty-four final spring sightings in Nebraska are from April 14 to May 22, with a median of May 6. Seventeen initial fall sightings are from September 5 to November 11, with a median of October 8. Seventeen final fall sightings are from October 9 to November 27, with a median of November 11. A state-level analysis of four decade-long periods of Christmas Bird Counts (1967-68 to 2006–7) extending from North Dakota to the Texas panhandle indicated a late-December population peak in Oklahoma (Johnsgard and Shane, 2009).

Habitats: Rivers, lakes and reservoirs while on migration. Breeding occurs on ponds and marshes ranging in size from less than an acre to several hundred acres, which may be seasonal or permanent. Submerged aquatic vegetation is typically abundant, but emergent growth may be rather sparse.

Breeding Status: Breeding occurs uncommonly or locally in north-central South Dakota. There were no reports of confirmed nestings during breeding bird atlas surveys in South Dakota (Peterson, 1995), Nebraska (Mollhoff, 2001) or Kansas (Busby & Zimmerman, 2001). There are fairly recent South Dakota breeding records from Dewey, Edmunds and McPherson counties, and historic records as far south as Beadle, Miner and Bennett counties (Tallman, Swanson and Palmer, 2002).

Breeding Habitats: Nesting in North Dakota occurs on fresh to slightly brackish wetlands that range from seasonal to permanent and vary in size from 1/2 acre to several hundred acres. Typically, abundant growths of submerged aquatic plants are present in breeding areas, but emergent vegetation is usually relatively sparse.

Nest Location: These birds are typically distributed as single pairs on ponds or widely scattered pairs on larger lakes or marshes, but as many as five nests have been located on a pond of 43 acres. The nest is usually built over dense beds of submerged vegetation, either in open water or in emergent vegetation near open water, with water depths varying from six to 48 inches.

Clutch Size and Incubation Period: From 3–6 eggs, usually four or five (13 clutches in North Dakota averaged 4.5). The eggs are white initially but become stained with brown. The incubation lasts 24–25 days and is probably by both parents. Typically the bird not incubating remains near the nest.

Time of Breeding: Nests in South Dakota have been reported from May 29 to June 24 (Peterson, 1995).

Breeding Biology: In North Dakota, horned grebes tend to select ponds that are fairly small (less than 2.5 acres) and contain mostly open water, which is evidently related to the importance of visual cues in territorial behavior. Displays are mostly mutual and include head-shaking, bill-touching ceremonies, weed-presentation ceremonies, standing vertically in the water facing the mate, and rushing over the water, often carrying vegetation in the bill. The "discovery ceremony" begins as one bird approaches another in a series of brief dives, then rises up ("ghostly penguin' display) facing away from the partner, who has waited in a hunched ("cat") posture. Both of the birds then move together in a prolonged "penguin dance" (Fjeldsa, 2004). The nest is built by both sexes, and copulation occurs on the nest platform. The young are initially tended and fed by both parents and often ride on their backs. After a few weeks one of the parents may leave the area while the other remains with the chicks. The fledging period is 41–50 days.

Comments: Slightly larger than the eared grebe, this is much less common, and lacks that species' black neck. There are no modern nesting records for this attractive species for Nebraska. National Breeding Bird Surveys between 1966 and 2009 indicate that the species underwent a statistically significant population decline (2.5 percent annually) during that period. The North American population may number more than 100,000 birds (Rose and Scott, 2000).

Suggested Reading: Storer, 1969; Fjeldsa 1973; 2004; Faaborg 1976, Ferguson and Sealey, 1983; Johnsgard, 1987; B.O.N.A. 505.

Red-necked Grebe, *Podiceps grisegena*

An extremely rare spring and fall migrant in Nebraska. There have been sightings of this species in at least six counties, with most of the reports for Douglas–Sarpy, and nearly all from counties bordering the Platte or Missouri rivers. It is s rare migrant in Kansas, and a rare to uncommon migrant in eastern South Dakota, breeding uncommonly and locally in the northeastern corner of the state.

Migration: Seven spring Nebraska records are from March 13 to May 17, with a mean of April 9, and eight fall Nebraska records are from September 30 to December 21, with a mean of October 30. Sharpe, Silcock and Jorgensen (2001) listed seven undocumented spring reports, from April 7 to May 13, and 13 documented plus ten undocumented fall reports, from September 26 to January 4, with a peak from October 24 to November 24. It was observed three years during Christmas Bird Counts at Lake McConaughy between 2000–2001 and 2009–2010.

Habitats: Rivers, lakes and reservoirs while on migration.

Breeding Status: Restricted during the breeding season to South Dakota, with breeding-season records from five counties (Tallman, Swanson and Palmer, 2002). There was one confirmed nestings during breeding bird atlas surveys in South Dakota (Peterson, 1995), but none in Nebraska (Mollhoff, 2001) or Kansas (Busby & Zimmerman, 2001).

Breeding Habitats: In North Dakota, this grebe nests on freshwater or slightly brackish permanent wetlands, usually at least ten acres in area. It also occurs on shallow river impoundments. Submerged plants, such as pondweeds, are usually present.

Nest Location: Red-necked grebes are in general solitary nesters, but "loose colonies" have been reported at a few Minnesota lakes. Pairs are usually well scattered on larger ponds and lakes, and their nests are either on open water or along the edges of emergent vegetation near open water. The water in such locations may be a foot or two in depth, and the nest is typically floating but anchored to vegetation. The nest is often constructed of submerged aquatic plants, and the location is probably chosen by the male. Both sexes participate in building the nest, with the male performing most of the work, and typically a single pair may construct several nests before one is completed and used.

Clutch Size and Incubation Period: Usually 3–5 eggs (seven North Dakota clutches averaged 3.1 eggs, and ranged from two to four). The eggs are initially white or pale bluish, but gradually become stained with brown. Incubation is by both sexes and lasts 22–23 days.

Time of Breeding: North Dakota egg dates range from May 19 to July 16, and dates of dependent young range from June 13 to August 22. Nests in South Dakota have been reported from May 6 to July 7 (Peterson, 1995).

Breeding Biology: Birds apparently arrive on their breeding grounds already paired. Territories are established that may include from 75–125 yards of shoreline, within which all breeding activities occur. Territorial and pair-forming displays are not yet well understood, but several mutual displays are performed by the pair. These include simultaneous calling while swimming side by side and erecting the crest, "rising breast to breast in the water, and emerging together from a dive, followed by rising in the water and facing each other, sometimes with vegetation in the bill, in a weed ceremony. Copulation occurs on the nest platform, following an invitation posture by the female. After hatching, both sexes brood and feed the young. The young remain on the nest for the first day, then ride on the backs of the parents for the first three weeks of life. They are fed for at least seven weeks,

and family bonds break up after eight to ten weeks. The fledging period is probably 7–9 weeks.

Comments: The red-necked grebe is a large, robust species that seems to prefer rather large bodies of water, and has a sharp bill well adapted to fish-catching. Like the eared grebe, it also occurs widely in Europe. The North American population is probably in excess of 100,000 birds (Rose and Scott, 2000).

Suggested Reading: Palmer 1962; Chamberlain 1977, Johnsgard, 1987; B.O.N.A. 465; Fjeldsa, 2004.

Eared Grebe, *Podiceps nigricollis*

An uncommon spring and fall migrant throughout Nebraska, and a fairly common summer resident, especially in the Sandhills, but locally also in the Rainwater Basin south of the Platte River. It also breeds in both Dakotas, and infrequently in Kansas.

Migration: A total of 105 initial spring sightings in Nebraska range from February 19 to June 5, with a median of April 22. Half of the sightings fall within the period of April 11 to May 5. Twenty-three final fall sightings are from August 23 to November 15, with a median of October 16. Half of the sightings fall within the period October 8–30. A state-level analysis of four decade-long periods of Christmas Bird Counts (1967-68 to 2006–7) extending from North Dakota to the Texas panhandle indicated a late-December population peak in Oklahoma (Johnsgard and Shane, 2009).

Habitats: Rivers, lakes and reservoirs during migration. Breeding occurs on ponds, marshes, and shallow river impoundments that are usually rich in submerged aquatic plants. Large, open ponds providing abundant feeding areas but also some sheltered locations with emergent aquatic plants for nesting sites seem to be favored.

Breeding Status: Breeds over nearly all of South Dakota and the Sandhills of Nebraska. There were 16 confirmed nestings during breeding bird atlas surveys in South Dakota (Peterson, 1995), ten in Nebraska (Mollhoff, 2001) and one in Kansas (Busby & Zimmerman, 2001). There are breeding records from three Kansas counties (Thompson, 2011).

Breeding Habitats: In North Dakota, eared grebes breed on wetlands that vary from slightly brackish to sub-saline, and from seasonal to permanent. They also use shallow river impoundments, and most nesting areas have extensive beds of submerged aquatic plants. Compared with horned and pied-billed grebes, eared grebes prefer larger, more open ponds that provide abundant feeding areas but also offer a sheltered location where a colony of nests can be placed.

Nest Location: Nests are generally less than 100 yards from the nearest shore and may be in open water or in emergent cover ranging from sparse to dense. Often nests in colonies may be separated by as little as ten feet, or in extreme cases may even touch each other. In seven colonies, the water depth ranged from as little as four inches to as much as 48 inches.

Clutch Size and Incubation Period: From 2–8 eggs are typical (101 nests in North Dakota averaged 3.8). The eggs vary from whitish to greenish or buffy. The incubation period is 20 1/2 to 21 1/2 days, and incubation is by both sexes. Apparently only one brood per season.

Time of Breeding: Egg dates in North Dakota range from May 21 to August 9, and extreme dates for dependent young are from June 8 to September 2. Nests in South Dakota have been reported from May 25 to July 7 (Peterson, 1995). In Nebraska, nests with eggs have been seen as early as June 2 (Mollhoff, 2001).

Breeding Biology: Pair-forming displays occur during spring migration while the birds are in flocks but continue after arrival on the breeding grounds. Courting occurs in the center of the breeding areas, and no territorial behavior is evident. Displays are mutual and include an advertising call by unpaired or separated birds, "habit-preening," head-shaking and a "penguin-dance" by both members of a pair standing upright in the water facing each other, and a "cat-attitude" with withdrawn head and fluffed body feathers. The female builds the nest, and copulation occurs on the nest platform, without elaborate associated displays. The nest is abandoned when the last egg hatches, and thereafter the young are tended by both parents, often riding on their backs. Young are relatively independent by their third week, but the actual fledging period is still unknown.

Comments: This beautiful little grebe is a colonial nester, and one of the summer attractions of Crescent Lake National Wildlife Refuge during mid-June is watching families of eared grebes with several chicks on the backs of nearly all adults. National Breeding Bird Surveys between 1966 and 2009 indicate that the species underwent a statistically non-significant population decline (0.3 percent annually) during that period. Scattered nestings have occurred in states to the south of Nebraska, including Kansas, Oklahoma, Texas and eastern Colorado. The North American population may number about five million birds (Fjeldsa, 2004).

Suggested Reading: McAllister 1958, Palmer 1962; Nuechterlein, 1975; 1981a; 1981b; Johnsgard, 1987; B.O.N.A. 433; Fjeldsa, 2004.

Western Grebe, *Aechmophorus occidentalis*

A common spring and fall migrant in the western parts of the region, rarer eastwardly, and a summer resident in western areas of Nebraska and South Dakota. Breeding occurs primarily on the larger Sandhills marshes, including Crescent Lake and Valentine refuges, which represent the southeastern limit of breeding of this species in the Great Plains. Many apparent non-breeders and a few breeders occur during summer on Lake Mc-Conaughy. There were two confirmed nestings during breeding bird atlas surveys in Kansas (Busby & Zimmerman, 2001), and many in South Dakota. It also breeds in North Dakota and southwestern Minnesota.

Migration: Seventy-seven initial spring sightings in Nebraska range from March 10 to June 10, with a median of May 6. Half of the sightings fall within the period April 19 to May 18. Forty-three final fall sightings are from September 10 to December 7, with a median of October 3. Half of the sightings fall within the period October 1 –24. As many as 44,000 have been seen on Lake McConaughy during September (Brown, Dinsmore, & Jorgensen, 2012).

Habitats: Rivers, lakes and reservoirs while on migration. Breeding is on ponds and lakes that usually have large expanses of open water, and on some marshes that are at least 50 acres in area.

Breeding Status: Breeds throughout most of South Dakota except the extreme southeast, the northern and western parts of Nebraska, and locally in central and western Kansas. There were 16 confirmed nestings during breeding bird atlas surveys in South Dakota (Peterson, 1995), three in Nebraska (Mollhoff, 2001) and two in Kansas (Busby & Zimmerman, 2001).

Breeding Habitats: In North Dakota, breeding occurs on permanent ponds and lakes that vary from slightly brackish to brackish and that contain large expanses of open water. Breeding also occurs on semi-permanent wetlands but usually is restricted to ponds of at least 50 acres. There were 16 confirmed nestings during breeding bird atlas surveys in South Dakota (Peterson, 1995), three in Nebraska (Mollhoff, 2001) and two in Kansas (Busby & Zimmerman, 2001).

Nest Location: Nesting is colonial, with nests at times numbering in the hundreds and birds in the thousands. In North Dakota, colonies have been found in areas where the water is from two to four feet deep in semi-open growth of emergent vegetation, usually hardstem bulrush or phragmites. In other areas cattails are sometimes used. Nests are often very close to one another; in a colony on Sweetwater Lake in North Dakota the average distance between nests was only about two yards. Sites offering protection from wave action and deep enough to allow underwater access to the nest are preferentially used.

Clutch Size and Incubation Period: Clutches range from three to seven eggs (12 North Dakota nests averaged 4.2). The eggs are very pale bluish green or buff initially, but soon become stained with brown. The incubation lasts 22–23 days, probably starting with the first egg laid, and is performed by both sexes.

Time of Breeding: Egg dates in North Dakota range from May 15 to June 10. Dates of dependent young range from June 13 to October 5. Nests in South Dakota have been reported from May 25 to July 16 (Peterson, 1995). In Nebraska, nests with eggs have been seen as early as June 6, and young as early as June 26 (Mollhoff, 2001).

Breeding Biology: Territorial activity by pairs is maintained only in the immediate vicinity of the nest, and most display activity occurs before the start of nesting, apparently serving primarily for pair-bond formation and maintenance. Most or all displays are performed by both sexes, often mutually. They include crest-raising while the birds swim together, with associated whistling notes and occasional withdrawal of the head and neck to the back, a "high arch" posture with neck stretched and bill pointed downward and the tail raised, a similar but less extreme "low arch" posture, ritualized "bob preening", and the "rush." In this last-named display, two birds (sometimes one and sometimes as many as six) call, then rise in the water and race side by side over the water surface with arched necks, bills pointed diagonally upward, and wings partially raised. Behavior leading to the race display usually includes threat-pointing with the bill and mutual bill-dipping as the birds approach; diving often terminates the display. When more than two birds perform the rush the additional birds are always males. Pair formation involves both the rush and "weed dancing," in which the two birds dive for weeds and then rise up nearly out of the water, raise their bills high, and rotate their heads from side to side, The ceremony ends when the weeds are thrown away and the pair performs bob-preening. (Fjeldsa, 2004). Copulation normally occurs on the nest site, but it has been observed on the edge of a beach, where the nests were on dry land. The fledging period is probably less than 70 days, by which time the flight feathers are fully grown.

Comments: The western grebe is perhaps the most spectacular of all the North American grebes, and during April as many as 14,000 of these splendid birds aggregate on Lake McConaughy. Breeding is common on several large marshes at Crescent Lake and Valentine refuges. National Breeding Bird Surveys between 1966 and 2009 indicate that the combined western & Clark's grebes underwent a statistically non-significant population decline (1.4 percent annually) during that period. The wonderful courtship displays of these grebes are perhaps best seen during May. The North American population may number 70,000 to 100,000 birds (Fjeldsa, 2004).

Suggested Reading: Palmer 1962; Nuechterlein 1975, Ratti, 1979; Storer and Nuechterlein. 1985; Johnsgard, 1987; B.O.N.A. 26; Fjeldsa, 2004.

Clark's Grebe, *Aechmophorus clarkii*

Occasional; probably a regular but local breeder in Nebraska and South Dakota. The state's first record was of an adult in breeding plumage found dead at Lake Keystone, Keith County on June 1, 1986, and since then a considerable number of sightings have been reported in western Nebraska, including adults with young at the western end of Lake McConaughy in late July, 1993 (Rosche, 1994). Up to 16 birds have been reported there, and some have also been seen at Crescent Lake National Wildlife Refuge. Sightings have occurred east to Lancaster County. In 1998 and 2002 breeding was noted at Willy Lake, Sheridan County. Breeding in South Dakota was first documented in 1987, and has been documented for Brown, Day, Kingsbury and Roberts counties. There was one confirmed nesting during breeding bird atlas surveys in South Dakota (Peterson, 1995). It is a rare migrant in Kansas, with no confirmed breeding records (Thompson, et al., 2011).

Breeding Status: Breeds very locally in Nebraska and South Dakota. There are no certain nesting records from Kansas. There was one confirmed nesting during breeding bird atlas surveys in South Dakota (Peterson, 1995), but none in Nebraska (Mollhoff, 2001) or Kansas (Busby & Zimmerman, 2001).

Habitats: Almost identical with that of the western grebe. The Clark's is said to prefer more open and deeper parks of lakes than the western (Fjeldsa, 2004).

Nest Location: Seemingly identical with that of the western grebe.

Clutch Size and Incubation Period: Seemingly identical with that of the western grebe.

Time of Breeding: Seemingly identical with that of the western grebe. A nest in South Dakota was reported on June 30 (Peterson, 1995).

Breeding Biology: Nearly identical with that of the western grebe, but having slightly different vocalizations (a single-note advertising call in the Clark's, and a double-note call in the western). These vocal differences, together with bill color differences and facial pattern differences, probably serve to reduce mixed-species matings. However, hybrids sometimes occur, and are known to be fertile. Males of the Clark's grebe often engages in the rushing ceremony, as do western grebes.

Comments: Probably this poorly distinguished species is more common in Nebraska than generally appreciated, since is very easily overlooked among flocks of western grebes. Mixed pairs and apparent hybrids have also been seen in the region (Thompson, et al., 2011; Brown, Dinsmore, & Jorgensen, 2012). Wintering birds were first seen in Nebraska during 1997–98 (*Nebraska Bird Review* 66:19). The breeding distribution of the Clark's grebe largely overlaps with that of the western grebe in the Great Plains, but the Clark's has not yet been proven to breed in Kansas. Population numbers are not, known, but in two cases of fairly large counts, the Clark's made up 11.6% and 49.2% of the mixed population.

Suggested Reading: Johnsgard, 1987; Ratti, 1979; Storer and Nuechterlein. 1985; B.O.N.A. 26; Fjeldsa, 2004.

ORDER PELECANIFORMES – PELICANS & CORMORANTS

Family Phalacrocoracidae–Cormorants

Neotropic Cormorant, *Phalacrocorax brasilianus*

Very rare throughout the three-state region, with at least 11 Nebraska re-cords through 2010 (*Nebraska Bird Review* 66:33; 67:72; 68:108, 78:135). It has also been reported from at least 14 central and eastern Kansas counties, and has bred in Barton County (Thompson *et al*, 2011). It has been observed at least twice in South Dakota (Tallman, Swanson and Palmer, 2002). Sharpe, Silcock and Jorgensen (2001) listed five documented and one undocumented reports, from March 26 to October 2.

Migration: In Oklahoma the records extend from February 27 to October 31, but the birds are usually present from April to October.

Habitats: Freshwater lakes and other wetlands, especially those with fish populations.

Breeding Status: There is a single Kansas breeding record for Cheyenne Bottoms Wildlife Area in 2007 (Thompson *et al.*, 2010). Breeds occasionally in northern Oklahoma, and regularly along the coast of Texas.

Breeding Habitats: Typically found in the same lake and pond habitats as those used by double-crested cormorants, and often associating with them.

Time of Breeding: The only known Kansas breeding occurred in mid-summer, with nestlings found on July 19, 2007.

Nest Location: The Kansas nesting occurred among a colony of double-crested cormorants. Generally the nest is in trees or bushes standing in water, from three to 20 feet above the substrate. Some nests are built on bare ground or rocks. The nests are bulky structures of twigs, and lined with coarse grass and stems.

Clutch Size: The clutch is typically of four eggs, ranging from three to six.

Breeding Biology: Nesting biology in this species is little-studied, but the incubation period has been reported as 24–25 days, and the period of dependency on the adults at about 84 days.

Comments: This species is clearly expanding its range northward at the present time, and future nestings are likely in the three-state region. Nesting already occurs as far north as northern Oklahoma. The total North American population is unknown. An early estimate of 8,7000 (Johnsgard. 1987) is probably far below current numbers.

Suggested Reading: Johnsgard, 1987; 1993; B.O.N.A. 137.

Double-crested Cormorant, *Phalacrocorax auritus*

An uncommon spring and fall migrant throughout three three-state region. It is a summer resident in several locations in the western half of Nebraska, east to Cherry County and the vicinity of North Platte. It also breeds locally in Kansas and South Dakota.

Migration: Of 102 initial spring sightings, the range is March 14 to May 29, and the median is April 12. Half of the records fall within the period April 14–25. Thirty-nine final spring Nebraska records range from April 17 to June 2, with a median of May 1. Thirty-one initial fall sightings are from August 7 to October 20, with a median of September 21. Thirty-one final fall sightings are from September 17 to December 14, with a median of October 23. A state-level analysis of four decade-long periods of Christmas Bird Counts (1967-68 to 2006–7) extending from North Dakota to the Texas panhandle indicated a late-December population peak in Oklahoma (Johnsgard and Shane, 2009).

Habitats: Migrating birds use deeper marshes, lakes, rivers and reservoirs. Breeding occurs on islands, trees, or cliffs near water, and within about ten miles of an adequate fish supply.

Breeding Status: Breeds in colonies scattered throughout much of the Dakotas and the western half of Nebraska. There were 36 confirmed nestings during breeding bird atlas surveys in South Dakota (Peterson, 1995), six in Nebraska (Mollhoff, 2001), and three in Kansas (Busby & Zimmerman, 2001). In Kansas, this species has bred (1951) at Cheyenne Bottoms Wildlife Area, Barton County, and since 1959 has nested at least periodically at Kirwin N.W.R., Phillips County.

Habitats: This species nests on rocky islands or cliffs adjoining water, or in trees in or near water. A supply of fish must be present within 5–10 miles from the nesting site. When trees are used, they may be either deciduous or coniferous, but they eventually die from the accumulation of excrement.

Nest Location: In North Dakota, nests in the Missouri River area are situated in the tops of dead trees, primarily cottonwoods, but colonies on islands of natural lakes are usually on the ground. One colony was also reported in willows that were in water 3 1/2 feet deep. Nests on solid substrates usually have a foundation of sticks, herbaceous vegetation, and rubbish, with finer materials added for lining. The male brings such material to the female, who incorporates it into the nest, and additional materials are added through the season. Garbage, including excreta and the remains of dead animals, also accumulates in the nest.

Clutch Size and Incubation Period: Normally from four to seven eggs (51 North Dakota nests averaged 4.6), but up to nine have been reported. The eggs are pale bluish with a chalky surface. The incubation period is 25–29 days. Flight might be attained by as early as 5–6 weeks of age. One brood per season.

Time of Breeding: Extreme egg dates for North Dakota are May 13 to July 18, and extreme dates of dependent young are May 31 to July 25. Nests in South Dakota have been reported from May 16 to mid-August (Peterson, 1995). Kansas egg records are for the period May 9 to June 20.

Breeding Biology: Cormorants are at least seasonally monogamous, usually breeding initially when three years old. Courtship occurs on water and includes much chasing and diving. Males choose the territory, which includes the nest and adjacent perching spot. Copulation occurs on the nest, mainly during the nest-building period. Both sexes assist in incubation, which begins before the clutch is complete; thus hatching is staggered over several days. The young leave the nest by about six weeks but continue to be fed by their parents until nine weeks of age, when family bonds disintegrate.

Comments: During the past few decades this cormorant has increased greatly in the U.S.A. benefiting from fish farms in the south and better protection on its nesting grounds. Nesting may occur in partly submerged bushes or trees, or on nesting platforms set out for Canada geese, as well as on sandy or gravely islands. National Breeding Bird Surveys between 1966 and 2009 indicate that the species underwent a statistically significant population increase (4.0 percent annually) during that period. The total North American population is unknown. A 1960–81 estimate of 220,000 (Johnsgard. 1993) is probably far below current numbers.

Suggested Reading: Palmer 1962; Mitchell 1977, Johnsgard, 1993; B.O.N.A. 441

Family Pelecanidae - Pelicans

American White Pelican, *Pelecanus erythrorhynchos*

A common migrant throughout the region. Non-breeders commonly occur through the summer on Nebraska's Harlan County Reservoir and Lake McConaughy, but there are no breeding records for the state. There were five confirmed nestings during breeding bird atlas surveys in South Dakota (Peterson, 1995), and none in Kansas (Busby & Zimmerman, 2001).

Migration: Eighty-four initial spring sightings in Nebraska range from February 21 to May 22 with a median of April 28. Half of the records fall within the period May 10–24. Thirty final spring sightings in Nebraska are

from April 12 to June 1, with a median of April 28. Twenty-eight initial fall sightings range from August 5 to November 21, with a median of September 24. Twenty-eight final fall sightings are from September 16 to November 10, with a median of October 16. A state-level analysis of four decade-long periods of Christmas Bird Counts (1967-68 to 2006–7) extending from North Dakota to the Texas panhandle indicated a late-December population peak in Oklahoma (Johnsgard and Shane, 2009).

Habitats: Deeper marshes, lakes and reservoirs are used by migrating and non-breeding birds. Breeding typically occurs on isolated and sparsely vegetated islands in lakes or reservoirs. For nesting, birds prefer islands that are nearly flat or have only gentle slopes, that lack obstructions that might interfere with taking flight, and that have loose earth easily worked into nest mounds. There were five confirmed nestings during breeding bird atlas surveys in South Dakota (Peterson, 1995), but none in Nebraska (Mollhoff, 2001) or Kansas (Busby & Zimmerman, 2001).

Breeding Status: Breeds in colonies in South Dakota (Waubay, Sand Lake, and Lacreek N.W.R.). Non-breeding birds are frequent during summer throughout most of the region, especially on large lakes or reservoirs.

Nest Location: Unlike some other pelican species, the white pelican nests only on the ground. A variety of nest materials have reportedly been used in nest construction, including shells, vegetation, dirt, sand, and stones.

Clutch Size and Incubation Period: From one to four eggs, usually two. The eggs are dull white, with a coarse texture. The incubation period is about a month (up to 36 days reported); incubation is by both sexes.

Time of Breeding: Extreme egg dates for North Dakota are June I to June 30; dates of dependent young range from June 1 to July 28. Nests in South Dakota have been reported from April 10 to August 19 (Peterson, 1995).

Breeding Biology: Pelicans are at least seasonally monogamous, and little display activity occurs on the nesting areas. Territorial defense is limited to the area immediately around the nest site, and most described displays occur at or near the nest. These include a "head-up" display with inflated or expanded gular pouch, which may serve as a greeting display to the mate and threat toward others, a "bow," with the bill pointed toward the feet and waved from side to side, and a "strutting walk" with the male following the female. Copulation occurs on land and is preceded by wing-quivering and squatting by the female. Both sexes incubate and share in feeding the young, but they feed only their own chicks. At the age of 50–60 days the young of the colony form a large "pod," and they fledge at about 10-11 weeks.

Comments: A major regional breeding site for this species is at Lacreek N.W.R. near Martin, South Dakota, but colonies also are known from Roberts, Marshall, Day and Codington counties, and has bred at Sand Lake N.W.R. in Brown County, when water levels have allowed islands to form. Probably many of the birds seen in Nebraska and Kansas are migrants going to and from South Dakota nesting sites or others farther north, or are non-breeders spending their summers away from the crowded breeding colonies, as occurs on Nebraska reservoirs such as McConaughy, Ogallala, and Calamus. National Breeding Bird Surveys between 1966 and 2009 indicate that the species underwent a statistically significant population increase (3.7 percent annually) during that period. The total North American population is unknown. A 1960–81 population estimate of 109,000 (Johnsgard. 1993) is probably far below current numbers.

Suggested Reading: Palmer 1962; Schaller 1964, Johnsgard, 1993; B.O.N.A. 57

ORDER CICONIIFORMES – HERONS, EGRETS & STORKS

Family Ardeidae - Herons, Bitterns & Egrets

American Bittern, *Botaurus lentiginosus*

A common spring and fall migrant throughout the three-state region, and a locally common summer resident. It breeds throughout Nebraska in suitable habitats, with the Sandhills marshes providing optimum habitat. It also breeds widely elsewhere in the region, except in the driest areas. There were three confirmed nestings during breeding bird atlas surveys in South Dakota (Peterson, 1995), and one in Kansas (Busby & Zimmerman, 2001).

Migration: The range of 109 initial spring sightings in Nebraska is from March 26 to June 10, with a median of May 3. Half of the records fall within the period April 23 to May 11. Forty-four final fall sightings are from July 14 to December 17, with a median of October 6. Half of the sightings fall within the period October 1–27.

Habitats: Normally this species is found in marshes, swamps and bogs having heavy emergent vegetation or with adjacent wet swales or tall grassy meadows.

Breeding Status: Nearly pandemic through the region. Throughout the area its breeding distribution is local, associated with marshes or tall grasslands. There were three confirmed nestings during breeding bird atlas surveys in South Dakota (Peterson, 1995), three in Nebraska (Mollhoff, 2001), and one in Kansas (Busby & Zimmerman, 2001).

Nest Location: The American bittern normally nests in a solitary manner in tall vegetation, usually cattails, bulrushes, or dense grasses, either on dry ground or on a mound several inches above the water. The nest platform is relatively scanty but usually is very well hidden from above and from the side, by the arching over of vegetation above it.

Clutch Size and Incubation Period: From two to five eggs (19 North Dakota nests averaged 3.9); eggs are buffy brown to olive buff, the surface smooth arid slightly glossy. The incubation period is 24–28 days. Single-brooded.

Time of Breeding: Egg dates in North Dakota range from May 31 to August 2. Nests in South Dakota have been reported from June 1 to August 12 (Peterson, 1995). In Kansas, courtship has been seen on April 29, egg records are for the period May 24 to June 27, and nestlings seen June 9–17. Oklahoma breeding dates range from May 5 (eggs) to August 5 (well-developed young).

Breeding Biology: Relatively little is known of the social behavior of this elusive bird, but males evidently establish and advertise territories with their distinctive "pumping" call, especially at dawn and dusk. However, the male starts no nest during this period. Females are attracted to such territories and form apparently monogamous (possibly polygamous) pair bonds. Copulation was once observed by the author to occur on open ground, after the male had displayed his white "shoulder" plumes and persistently advanced toward the female while repeatedly lowering and swaying his head from side to side, as if he were regurgitating food. After overtaking the retreating female he simply climbed on her back, grasped her nape, and copulated. No specific postcopulatory behavior was noted. The female evidently chooses the nest location (about 50 yards from the site of copulation in a case personally observed), and apparently does all the nest-building and incubation. The male takes no part in defending the nest, but the female defends it fiercely. The young remain in the nest for about two weeks, but the fledging period is evidently still unknown. In the closely related Eurasian species (*B. stellaris*) the period is 55–60 days, which probably closely approximates that of the American bittern.

Comments: This strange heron, usually disparagingly called a "shitepoke" by native Nebraskans, often goes unseen by casual observers, who fail to notice it standing erect and motionless among cattails and reeds near water. Its booming courtship call is responsible for its alternative vernacular name, "thunder-pump." National Breeding Bird Surveys between 1966 and 2009 indicate that the species had a statistically non-significant population decline (1.0 percent annually) during that period.

Suggested Reading: Mousley 1939; Palmer 1962; Hancock and Elliott, 1978, Hancock and Kushlan, 1984; B.O.N.A. 18; Kushlan and Hancock 2005; Johnson & Igle, varied dates.

Least Bittern, *Ixobrychus exilis*

An uncommon spring and fall migrant and seemingly rare summer resident in eastern parts of the three-state region. It breeds locally in the eastern half of Nebraska, and perhaps has its western nesting limits in Garden County, where it has been observed during summer at the Ash Hollow marshes and Crescent Lake National Wildlife Refuge. There were four confirmed nestings during breeding bird atlas surveys in South Dakota (Peterson, 1995), where it occurs mainly east of the Missouri River, and one in Kansas (Busby & Zimmerman, 2001), where it is most common in eastern and central parts of the state.

Migration: Thirty-nine initial spring sightings in Nebraska range from March 30 to June 4, with a median of May 15. Half of the records fall within the period May 4–24. Ten final fall sightings are from July 28 to September 19, with a median of August 17.

Habitats: In this region the least bittern is associated with freshwater or slightly brackish marshes and lakes that have extensive stands of cattails, bulrushes, and other rank vegetation. It is thus not usually found around large impoundments or rivers, where water levels may fluctuate. Marshes with scattered bushes or similar woody growth are favored.

Breeding Status: Probably breeds over the eastern half of the entire region under consideration, but apparently fairly uncommon to rare throughout. It is relatively rare in Nebraska, judging from the few nesting records. There are nesting records in South Dakota for Charles Mix, Day, Deuel, Lincoln, Sanborn and Yankton counties (Tallman, Swanson and Palmer, 2002), and for nine Kansas counties, from Barton and Stafford counties eastward (Thompson *et al.*, 2011). There was one confirmed nesting during breeding bird atlas surveys in South Dakota (Peterson, 1995), three in Nebraska (Mollhoff, 2001), and four in Kansas (Busby & Zimmerman, 2001).

Nest Location: Nests are built above shallow water, in living or partly living stands of bulrushes or cattails, often close to open water. The nest is made of both dead and living materials and is usually about 6–8 inches across, round or oval. It has a foundation of dried leafy material and twigs that are arranged in a spoke-like manner a foot or two above the water, with arched over vegetation above.

Clutch Size and Incubation Period: From three to six eggs; the clutch is often about five in northern areas and somewhat smaller farther south. In Iowa, 59 clutches ranged from two to six eggs and averaged 4.4. The eggs are very pale bluish or greenish with a smooth surface. The incubation period is 17–18 days. Regularly double-brooded, at least in some areas or in favorable years.

Time of Breeding: Nests in South Dakota have been reported from June 3 to July 13 (Peterson, 1995). In Nebraska, nests with eggs have been seen as early as June 16, and young as early as July 7 (Mollhoff, 2001). Kansas egg dates range from June 7 to July 11, and nestlings seen from June 20 to July 14.

Breeding Biology: In Iowa, birds arrive on their breeding marshes about two weeks before the start of nesting. Little information is available on territoriality, but since nests are often fairly close to one another, territories must be rather small. The male's advertising call is a series of soft cooing notes, and presumably visual displays are also performed. Males evidently choose a nest site and do the early nest-building, as in other herons. Pair-forming displays still are unknown but probably involve mutual preening and crest-raising, since these occur during nest-relief ceremonies. Both sexes incubate, sharing incubation time about equally. Likewise, both sexes feed the young, but the male assumes the major role in this. The young usually remain in the nest for about 10–14 days but may leave it for short periods when only six days old. The adults continue to feed the young after they have left the nest, but at least at times they soon begin a second clutch.

Comments: This is the smallest of the American herons, and a miniature version of the American bittern. It builds a distinctive nest of materials organized in a radiating configuration in marshy vegetation. National Breeding Bird Surveys between 1966 and 2009 indicate that the species had a significantly stable population (0.0 percent change annually) during that period.

Suggested Reading: Weller 1961; Palmer 1962; Hancock and Elliott, 1978; Hancock and Kushlan, 1984; B.O.N.A. 17; Kushlan and Hancock 2005.

Great Blue Heron, *Ardea herodias*

A common migrant and a local summer resident throughout the three-state region, breeding in colonies in various locations throughout, but especially along tree-lined rivers and larger wetlands. It also breeds locally elsewhere throughout South Dakota and Kansas. There were 50 confirmed nestings during breeding bird atlas surveys in South Dakota (Peterson, 1995), and 63 in Kansas (Busby & Zimmerman, 2001).

Migration: The range of 87 initial spring sightings in Nebraska is from January 6 to June 6, with a median of April 2. Half of the records fall within the period March 26–April 30. Of 103 final fall sightings, the range is August 8 to December 30, and the median is October 13. Half of the records fall within the period September 23–November 7. A state-level analysis of four decade-long periods of Christmas Bird Counts (1967-68 to 2006–7) extending from North Dakota to the Texas panhandle indicated a late-December population peak in Oklahoma (Johnsgard and Shane, 2009).

Habitats: Migrants are found around all wetlands and streams supporting a fish population and having shallows for foraging. Nesting usually occurs among groves of tall trees, but sometimes also has been reported on the ground, on rock ledges, among bulrushes, or other elevated situations. Cottonwood groves seem to be a favored nesting location in Nebraska. Within the Great Plains region, herons often nest in association with cormorants, especially where reservoirs have flooded tall trees.

Breeding Status: Pandemic and common throughout the region, breeding locally along many rivers, lakes, and reservoirs. There are breeding records for all but 16 Kansas counties (Thompson *et al.,* 2011). There were 50 confirmed nestings during breeding bird atlas surveys in South Dakota (Peterson, 1995), 18 in Nebraska (Mollhoff, 2001), and 63 in Kansas (Busby & Zimmerman, 2001).

Nest Location: Nests are usually placed in a crotch or on a large limb of a tall tree, sometimes more than 100 feet above the ground. Usually more than one nest occurs per tree in large colonies, and old nests are frequently reused. Nests that have been used for several years tend to be massive; newly made ones are often flimsy. Adults continue to add materials to the nest until the young are well grown.

Clutch Size and Incubation Period: From three to six eggs (36 Kansas clutches averaged 4.4), pale bluish green, smooth to slightly rough in texture. The incubation period is 25–29 days. One brood per season.

Time of Breeding: Egg dates in North Dakota range from April 27 to May 15. Nests in South Dakota have been reported from May 28 to July 29 (Peterson, 1995). In Nebraska, nests with eggs have been seen as early as May 19 (Mollhoff, 2001). Kansas egg records are from March 15 to April 30, and young have been noted from May 28 to August 11..

Breeding Biology: Great blue herons are seasonally monogamous, and both sexes arrive at the nesting ground about the same time. Birds probably breed initially when two years old, but some variation is likely. The male selects the breeding territory, which usually centers on an old nest. Several obviously hostile displays are associated with territorial defense. Additionally, numerous highly ritualized territorial advertising displays occur, including the "stretch," "snap," and others. These are predominantly male displays, given at the nest site, and serve to attract females and aid pair-formation. Mutual behavior between members of a pair includes twig-passing, feather-nibbling, bill-stroking, and similar activities. Copulation is sometimes preceded by displays, such as feather-nibbling, but it may also occur without obvious display. When building or improving the nest, the male gathers materials and the female works them into the nest. Both sexes incubate, and nest-relief ceremonies are performed. The eggs typically hatch over an interval of 5–8 days, and adults feed the young by regurgitating food into the bottom of the nest. Although the young can make short flights in the nest vicinity when seven weeks old, they usually continue to use the nest and are fed by the adults until they are about 10–11 weeks old.

Comments: Commonly erroneously called "cranes," great blue herons differ from cranes in many ways, including their tree-nesting behavior, their flight profile (flying with kinked-necks rather than outstretched necks), and their strongly fish-dependent diet. However, they are beautiful and graceful birds, and often nest colonially in areas where fish are plentiful. National Breeding Bird Surveys between 1966 and 2009 indicate that the species underwent a statistically significant population increase (1.0 percent annually) during that period. There are no nationwide population estimates. But one 1980's survey estimated 36,000 breeding birds along the U.S. coast from Texas to Maine, plus the Great Lakes (B.O.N.A. 25).

Suggested Reading: Pratt 1970; Mock 1976; Hancock and Elliott, 1978; Hancock and Kushlan, 1984; B.O.N.A. 25: Kushlan and Hancock 2005.

Great Egret, *Ardea alba*

A common (Kansas) to uncommon or occasional (Nebraska and South Dakota) spring and fall migrant throughout the region, breeding locally and most frequently in Kansas. It is most common in the region's eastern counties, but has been observed as far west as at least Box Butte and Garden counties in Nebraska and the counties bordering Colorado in Kansas. There was an attempted nesting in Sarpy County in 1960 (*Nebraska Bird Review* 28:55), the only confirmed state nesting record (Sharpe *et al.*, 2001). There are South Dakota nesting records for at least Brown, Charles Mix, Codington, Day and Kingsbury counties, and large nesting colonies found in Brown and Clark counties (Tallman, Swanson and Palmer, 2002). It is a local but regular breeder in eastern Kansas, with nesting records for at least seven central and eastern counties as of 2011 (Thompson *et al.*, 2011). There were two confirmed nestings during breeding bird atlas surveys in South Dakota (Peterson, 1995), and four in Kansas (Busby & Zimmerman, 2001).

Migration: Sixty-two initial or only spring Nebraska records are from March 26 to June 1, with a median of April 29. Half of the records fall within the period April 16 to May 10. Ten final spring sightings in Nebraska are from April 6 to June 8, with a median of May 9. Twenty-one total fall Nebraska records are from August two to October 21, with a median of September 1. Out of 95 total records, the largest number (34) are for May, followed by April (30), August (10) and September (8).

Habitats: The species occurs in freshwater, brackish, and occasionally salt-water Habitats. It is found on streams, swamps, and lake borders, usually close to trees during the nesting season, but forages in fairly open situations.

Breeding Status: Breeds locally in Kansas and South Dakota. Only one breeding record is known for Nebraska, but summer visitors are fairly common, and local breeding can be expected. There were six confirmed nestings during breeding bird atlas surveys in South Dakota (Peterson, 1995), none in Nebraska (Mollhoff, 2001), and four in Kansas (Busby & Zimmerman, 2001).

Nest Location: Nests are either solitary or in colonies, usually with other species of herons, such as great blue herons. The nests are generally between ten and 30 feet above the ground in trees, frequently beeches or maples in the northern states and cypress in the south. They tend to be very high up and to be less bulky than those of great blue heron, but are larger than those of the smaller heron species.

Clutch Size and Incubation Period: From two to four eggs, but three eggs apparently are most common. The eggs are blue to greenish blue, with a smooth surface. The incubation period is 23–24 days. One brood per season.

Time of Breeding: Nests in South Dakota have been reported from June 16 to August 1 (Peterson, 1995). In Kansas, the breeding season extends from early April through June. In Oklahoma, nest-building has been reported from April 4 to May 20, eggs noted May 23, and dependent young seen July 2 to August 17.

Breeding Biology: In the first phase of breeding males establish territories that center on nest sites or old nests, preferably the latter, since these allow for earlier advertisement displays. When a nest platform is available, the males perform several courtship displays, including a ritualized preening movement or "wing-stroking," the "stretch" display (a vertical neck-stretch followed by a bobbing movement), the "bow" (a repeated twig-shoving movement followed by a bob), the "snap" (a downward extension of the head and neck, accompanied by a mandible snap and a bob), and a circular flight. Males thus attract females to the nest site, where copulation occurs. Within a few days a pair bond is formed, and shortly thereafter egg-laying begins. Both sexes incubate, and they perform greeting ceremonies when exchanging places on the nest. Likewise, both sexes feed the young, which require approximately six weeks to attain flight.

Comments: The range of several egrets seems to have expanded in recent years, and that is certainly the case with the great egret. There are influxes of the birds into Nebraska just prior to breeding (perhaps of birds headed toward Minnesota and western Iowa), and again following the breeding season. National Breeding Bird Surveys between 1966 and 2009 indicate that the species underwent a statistically significant population increase (2.6 percent annually) during that period.

Suggested Reading: Wiese 1976; Tomlinson 1976; Hancock and Elliott, 1978; Hancock and Kushlan, 1984; B.O.N.A. 570; Kushlan and Hancock, 2005.

Snowy Egret, *Egretta thula*

A common (Kansas) to uncommon (Nebraska and South Dakota) migrant or summer visitor throughout the three-state region. Most Nebraska occurrence records are for counties bordering the Platte or Missouri Rivers, especially Douglas-Sarpy, Lincoln, Platte, Scotts Bluff, and also Lancaster. There are breeding records from Hall, Lancaster and Scotts Bluff counties (Mollhoff, 2001). It regularly breeds in South Dakota, and also breeds locally in Kansas. Post-breeding northward dispersal is frequent as far north as South Dakota, bringing increased numbers into the region during July and August.

Migration: Twenty-four total spring Nebraska records are from April 13 to June 10, with a median of May 7. Ten total fall Nebraska records are from July 30 to October 1, with a median of August 17. Of 34 total records, the largest numbers are for May (17) followed by April (6) and August (4).

Habitats: Snowy egrets occupy habitats ranging from freshwater to saline but prefer relatively sheltered locations. Ponds with low willows, buttonrush, and similar shrubs are favored, as are thick stands of mangroves. In southern areas the birds are usually found in heronries of little blue herons, great egrets, and black-crowned night herons.

Breeding Status: There are South Dakota breeding records from Bennett, Brown, Clark, Charles Mix, Codington, Day and Kingsbury counties (Tallman, Swanson and Palmer, 2002). There were two confirmed nestings during breeding bird atlas surveys in South Dakota (Peterson, 1995), and one in Nebraska (Mollhoff, 2001). There are breeding records from six central and western Kansas counties (Thompson, 2011), and four nestings were confirmed during breeding bird atlas surveys in Kansas (Busby & Zimmerman, 2001).

Nest Location: Nests are usually in shrubs or low trees, from two to ten feet above the ground, but up to 30 feet has been recorded. The birds are typically colonial but may nest singly at the edge of their range. The nests are rather flat and elliptical rather than round and are loosely constructed. They are often built of slender twigs a foot or two long, gathered close to the nest site.

Clutch Size and Incubation Period: From two to five eggs, with four probably most typical. The eggs are pale greenish blue, with smooth shells. The incubation period is 22–23 days. One brood per season, but early nesting losses may be replaced.

Time of Breeding: Nests in South Dakota have been reported from May 9 to July 29 (Peterson, 1995). In Nebraska, adults with young have been seen as early as June 26 (Mollhoff, 2001). In Kansas, nest-building have been reported from April 21 to June 7, eggs from May 27 to July 11 (and also September 1), and nestlings June 7 to August 31.

Breeding Biology: After returning to their breeding grounds, males establish a territory that centers on a potential nest site but need not include an old nest. Besides hostile displays, the male performs several sexual displays that include both a stationary and an aerial "stretch" as major advertisement displays. A single "circle flight" around the potential mate is also common, and a more spectacular flight is a towering circular flight from 50 to 150 yards above the female, followed by a spectacular tumbling downward to land beside her. A mutual display called the "jumping over" display, in which one bird makes a short jump flight over the back of the other, is a probable indication that a pair bond has been formed. The male gathers material and the female constructs the nest. Copulation occurs on the nest site or on a limb close to it. The first egg may be laid before the nest is completed, and eggs are laid about two days apart. Since incubation (by both sexes) begins before the clutch is complete, the first young hatches about 18 days after the last egg is laid. After 20–25 days the young are ready to leave the nest, and they are independent by 7–8 weeks.

Comments: This beautiful little egret can be easily recognized by its "golden slippers" and its all-black bill. In the spring it exhibits beautiful filmy white "aigrette" plumes that once where the high-fashion rage and nearly spelled the species' doom before federal protected was enacted. National Breeding Bird Surveys between 1966 and 2009 indicate that the species underwent a statistically non-significant population increase (1.6 percent annually) during that period.

Suggested Reading: Mavericks 1960; Jenni, 1969; Hancock and Elliott, 1978; Hancock and Kushlan, 1984; B.O.N.A. 489; Kushlan and Hancock, 2005.

Little Blue Heron, *Egretta cerulean*

An uncommon (Kansas) occasional (Nebraska) or rare (South Dakota) spring and fall migrant, primarily in the eastern parts of the region. Breedings have been reported for South Dakota and Kansas, but not Nebraska. There are records from many Nebraska counties, but the largest numbers of sightings are for Adams, Platte, Lancaster, and Douglas–Sarpy. Summer visitors may be seen throughout most of the region.

Migration: A total of 55 spring Nebraska records range from April 4 to June 1, with a median of May 8. Fifteen fall Nebraska records are from July 23 to October 20, with a median of August 19. Of 80 total records, the largest numbers (37) for May, followed by April (16), June (8) and July (7).

Habitats: Although found both in freshwater and saline environments, this species is mostly limited to inland habitats such as woodland ponds.

Breeding Status: There were no confirmed nestings during breeding bird atlas surveys in South Dakota (Peterson, 1995), but nestings did occur in 1985 and 1995. There were seven confirmed nestings during breeding bird atlas surveys in Kansas (Busby & Zimmerman, 2001), but all Kansas breeding records have been confined to Barton and Stafford counties (Thompson *et al.*, 2011).

Nest Location: The species is colonial in nesting; nests are situated from a few feet above the ground or water to as much as 40 feet. In a Florida study, the birds usually nested on horizontal limbs with the nest wedged against the main trunk, at an average of about seven feet above the substrate. They were thus less exposed than snowy egret nests in the same area and tended to be slightly higher. In Oklahoma a variety of broad-leaved trees have been used for nesting, often shared with snowy and great egrets.

Clutch Size and Incubation Period: From three to six eggs (58 Florida clutches averaged 3.7). The eggs are pale greenish blue with a blue gloss. The incubation period is 22–25 days, averaging 22.8 days, and with staggered hatching of the young. One brood per season.

Time of Breeding: Oklahoma egg records range from April 19 to July 7, and observations of dependent young extend from May 20 to July 7.

Breeding Biology: Males begin to establish territories a few weeks before egg-laying by defending an area about 25 feet in diameter around an old nest or nest site. Besides various threat displays, the "stretch" display is perhaps the most important sexual display. Unmated females are attracted to such males but are initially repulsed. Besides the stretch display, the "snap" display, with mandible clicking, is common. Early stages of pair-formation including mutual billing, neck-crossing and intertwining, but virtually no aerial displays as in the green heron and snowy egret. A strong pair bond is formed, but some promiscuous copulatory behavior has been observed. Copulation occurs on the nest platform or close to it. The female completes the nest started by the male, and the male passes twigs to her in an elaborate ceremony. Little nest-building occurs after incubation gets under way;

both sexes participate equally in incubation. The young are fed by both parents, who regurgitate food into their mouths or into the nest. The young are able to make short flights when five weeks old.

Comments: National Breeding Bird Surveys between 1966 and 2009 indicate that the species underwent a statistically significant population decline (2.6 percent annually) during that period.

Suggested Reading: Meanley 1955; Palmer 1962; Hancock and Elliott, 1978; Hancock and Kushlan, 1984; B.O.N.A. 145; Kushlan and Hancock, 2005.

Tricolored Heron, *Egretta tricolor*

A rare summer visitor and very rare breeder in Kansas, and a very rare summer visitor and breeder in South Dakota. Accidental in Nebraska. There were only five Nebraska records through 2001, including one shot in Kearney County in October of 1918, another collected in Clay County in August 1918. There are sight records from Clay County in August, 1971 (*Nebraska Bird Review* 41:14), and from Hitchcock County in 2001 (*Nebraska Bird Review* 71:98). Sharpe, Silcock and Jorgensen (2001) listed two undocumented spring reports, May 4 and 22, and three summer or fall reports, including those mentioned above.

Migration: In Kansas, early dates are April 16 to May 27 in Barton County, and April 23 to May 21 in Stafford County. Most late dates are from September 4–27, with one as late as October 2 (Thompson *et al.*, 2011).

Habitats: During the breeding season this species is primarily found near salt water, inhabiting mangroves, tidal marshes, and similar habitats.

Breeding Status: There are 1974, 1976 and 1985 breeding records for Barton County, Kansas, and breeding efforts for at least 15 summers at Quivira NWR or Cheyenne Bottoms Wildlife Area since 1965 (Thompson *et al.*, 2011). The only South Dakota breeding records are from Brown County in 1995 (a hybrid mating with a snowy egret) and 2001, and two nests in Kingsbury County in 1986 (Tallman, Swanson and Palmer, 2002).

Nest Location: This species builds nests closer to the substrate than most other herons, usually less than seven feet up, and rarely above ten feet. The nests also tend to be in more sheltered and sturdier locations than those of snowy egrets. The species is highly social, at least in most areas.

Clutch Size and Incubation Period: From three to six eggs, usually four. The eggs are pale greenish blue with a smooth surface. The incubation period is 23–25 days, averaging about 24 days. One brood per season.

Time of Breeding: Kansas egg records are for the period June 20 to July 5.

Breeding Biology: As in the other herons, the male establishes a territory that includes a nesting site and displays within it, threatening other males and attracting unpaired females. Several threat displays are present, as well as various sexual displays. The most elaborate of these is a combined stretch and snap display, which includes sudden extension of the head and neck, seizing a twig and dropping it, and a series of strong pumping movements. Females are initially evicted from the territory but are gradually accepted, and soon the pair begins mutual nibbling and billing. The male builds the foundation of the nest before pair formation, but the female completes it while the male gathers material. Copulation occurs on the nest or beside it, before and probably during the egg-laying period. Both sexes incubate and care for the young, which hatch at intervals and remain in the nest about two weeks. As in other herons, many of the nestling losses result from starvation of the youngest chick. By the time they are 21–24 days old the young are fed away from the nest, and feathering is complete at about four weeks. They are independent at an average age of 59 days.

Comments: National Breeding Bird Surveys between 1966 and 2009 indicate that the species underwent a statistically non-significant population decline (0.2 percent annually) during that period. There may have been a minimum of about 97,000 pairs in the U.S during the 1970's (B.O.N.A. 306).

Suggested Reading: Jenni 1969; Rogers 1977; Hancock and Elliott, 1978; Hancock and Kushlan, 1984; B.O.N.A. 306; Kushlan and Hancock, 2005.

Cattle Egret, *Bubulcus ibis*

A common to uncommon spring and fall migrant throughout the region. It is a locally uncommon breeder in South Dakota and Kansas, and a rare Nebraska nester. It was first reported in Kansas and South Dakota in 1961, in Nebraska in 1965, and has been observed with increasing frequency since then. Most Nebraska records are for Douglas–Sarpy, Lancaster and Otoe counties.

Migration: Twenty-one total spring Nebraska records range from April 12 to June 3 with a median of May 9. Eleven total fall Nebraska records are from July 23 to October 20, with a median of August 29. Of 32 total records, the largest number (11) are for May, followed by April (8) and August (5).

Habitats: Cattle egrets occur in a wide variety of freshwater to saline habitats and are more terrestrial than any native North American herons. They are highly social and normally nest among other herons. They are usually found near cattle in North America and forage largely on grasshoppers and other insects rather than on fish like most herons.

Breeding Status: Nesting in Nebraska has been documented for at least four Nebraska counties: Cherry, Garfield, Holt and Keith (Sharpe et al., 2001). There were two confirmed nestings during breeding bird atlas surveys in South Dakota (Peterson, 1995), one in Nebraska (Mollhoff, 2001), and six in Kansas (Busby & Zimmerman, 2001). Nesting records exist for six South Dakota counties (Brown, Charles Mix, Corson, Douglas, Edmunds and Kingsbury (Tallman, Swanson and Palmer, 2002) and for five Kansas counties: Baton, Reno, Sedgwick, Seward and Stafford (Thompson *et al.*, 2011).

Nest Location: Compared with other small herons, cattle egrets tend to nest in relatively dense vegetation, at heights that are variable but usually under 20 feet, averaging about seven feet. At least in Florida, cattle egrets nest somewhat later than other herons, and their nests are more complete at the time of egg-laying.

Clutch Size and Incubation Period: From one to six eggs (85 Florida clutches averaged 3.5), very pale blue or bluish white with a smooth surface. The incubation period is 22–23 days, usually 23. The chicks hatch at intervals, usually two days apart. One brood per season.

Time of Breeding: Nests in South Dakota have been reported from June 1 to July 26 (Peterson, 1995). In Nebraska, nests with eggs have been seen as early as June 22 (Mollhoff, 2001). Kansas egg records are for the period June 5 to July 27. Dependent young have been observed in Oklahoma between June 30 and August 21.

Breeding Biology: The cattle egret maintains a smaller breeding territory than other heron species, which is related to its high degree of coloniality. Males establish territories that initially cover only a few square yards and are soon reduced to the immediate area around the nest. The male performs several threat displays within this territory, and he also performs several visual courtship displays ("stretch," "twig-shaking," "wing-touching," "forward-snap," "flap-flight," and "forward"), which are similar to those of other herons. Females are attracted to a displaying male and form a pair bond by flying to him, landing on his back, and subduing his aggressive tendencies by repeated blows on the head. These blows gradually change to nibbling after the male has ceased to fight back. Mutual back-biting is used thereafter by the pair as a greeting display, and it often precedes copulation. Some instances of polygamous pair bonds have been seen, but monogamy is the general pattern. The female completes the nest started by the male, which may require up to six days. Both sexes assist in incubation, with the female apparently sitting most of the daylight hours and the male at night. Compared with other herons, cattle egrets are very attentive to their young, and nestling mortality is low, compensating for their relatively small clutch size. The young begin to leave the nest at about 20 days of age, and fledge at about 30 days.

Comments: Since this species found its way to America from Africa, it has spread widely, Feeding on insects disturbed by foraging cattle on the American plains is seemingly little different from feeding around the feet of zebras and wildebeest on the savannas of East Africa. National Breeding Bird Surveys between 1966 and 2009 indicate that the species underwent a statistically non-significant population decline (0.6 percent annually) during that period.

Suggested Reading: Jenni 1969; Lancaster 1970; Hancock and Elliott, 1978; Hancock and Kushlan, 1984; B.O.N.A. 113; Kushlan and Hancock, 2005.

Green Heron, *Butorides virescens*

A common migrant, and a summer resident throughout the region, breeding over most of Nebraska and Kansas excepting the westernmost areas, and the eastern half of South Dakota.

Migrations: The range of 93 initial spring sightings in Nebraska is from March 10 to June 7, with a median of April 27. Half of the records fall within the period April 15 to May 6. The range of 50 final fall sightings is from July 23 to November 9, with a median of September 18. Half of the records fall within the period September 4–25.

Habitats: Migrating birds occur almost anywhere small fish (such as minnows) can be captured. This species occupies a broad range of habitats and water types, usually near trees, but also sometimes breeds in marshlands well away from tree cover. One of the most adaptable of North American herons, usually breeding as solitary pairs or in loose colonies.

Breeding Status: There was one confirmed nesting during breeding bird atlas surveys in South Dakota (Peterson, 1995), eight in Nebraska (Mollhoff, 2003) and 25 in Kansas (Busby & Zimmerman, 2001).

Nest Location: The nest is usually 10–15 feet above the ground, depending on the habitat, but may be directly on the ground or up to 30 feet above it. It varies in form from very flimsy to very bulky, the latter usually when it has been used many times. At times the old nests of other herons are also used, and the birds thus sometimes nest among other species of herons or egrets.

Clutch Size and Incubation Period: From three to six eggs, with four or five most common in northern part of range and fewer toward the south (17 Kansas clutches averaged 3.1). The eggs are pale greenish or bluish green, with smooth surface. The incubation period is 19–21 days, with incubation starting before the last egg is laid, resulting in a staggered period of hatching. At least in some regions two broods are produced per season, but there is no evidence of that in this region.

Time of Breeding: Nests in South Dakota have been reported from May 30 to July 7 (Peterson, 1995). In Nebraska, nests with eggs have been seen as early as May 29 (Mollhoff, 2001). In Kansas, egg records span the period April 21 to July 14.

Breeding Biology: Males select and defend territories on their return to the breeding grounds; separate feeding territories may also be defended. Initially quite large, the male's territory soon shrinks to the area around the nest or nest site. The territory is advertised by a "flying-around" display over the breeding site and by an advertising call from a conspicuous perch. Males also perform "stretch" and "snap" displays similar to those of the larger herons, and after a female has been attracted to the territory both sexes perform "circle-flight," "crooked-neck-flight," and "flap-flight" displays. After pair bonds have formed, the female completes the nest; the male helps in gathering materials. Copulation occurs on the nest platform or an adjacent branch and continues through egg-laying. Both sexes share in incubation and perform nest-relief ceremonies. The young hatch at intervals and are fed by regurgitation. They remain in the nest for about 16–17 days but do not actually fledge until they are about 21–23 days, with adults continuing to feed them until that time. In areas where two broods are raised, the second clutch may be begun only nine days after the first brood has fledged.

Comments: This widespread little heron is notable for the fact that it has been seen catching small minnows, disabling them, and then releasing them to serve as bait for attracting larger fish. National Breeding Bird Surveys between 1966 and 2009 indicate that the species underwent a statistically significant population decline (1.5 percent annually) during that period.

Suggested Reading: Meyeriecks 1960; Palmer 1962; Hancock and Elliott, 1978; Hancock and Kushlan, 1984; B.O.N.A. 129; Kushlan and Hancock, 2005.

Black-crowned Night-Heron, *Nycticorax nycticorax*

A common migrant throughout the region, breeding locally in suitable habitats throughout, except perhaps in westernmost areas.

Migration: Eighty initial spring sightings in Nebraska range from March 29 to June 9, with a median of April 25. Half of the records fall within the period April 18 to May 10. Fifty-four final fall sightings are from July 22 to November 15, with a median of September 6. Half of the records fall within the period August 18 to September 29.

Habitats: This is a highly adaptable species, found in a wide array of aquatic habitats, with nesting occurring in swamps, marshes, and sometimes even city parks or orchards where water is nearby.

Breeding Status: Breeds locally over most of the region. There were nine confirmed nestings during breeding bird atlas surveys in South Dakota (Peterson, 1995), five in Nebraska (Mollhoff, 2001) and 13 in Kansas (Busby & Zimmerman, 2001).

Nest Location: The species nests colonially, on dry ground, in bulrush or cattail marshes, or in trees up to 160 feet above the substrate. The nests are closely placed and often conspicuously situated. Newly made nests are flimsy, but they gain size and substance with repeated use. Nests are often situated in heronries that include other species. In our region, nesting is most frequent in bulrush or phragmites marshes or in groves of trees near rivers, often cottonwoods.

Clutch Size and Incubation Period: From two to six eggs; Kansas clutches are typically about four eggs. The eggs are pale blue or greenish blue with a smooth surface. The incubation period is 24–26 days; possibly double-brooding occurs in some areas.

Time of Breeding: Nests in South Dakota have been reported from May 16 to July 27 (Peterson, 1995). In Nebraska, nests with eggs have been seen as early as June 22, and young as early as July 20 (Mollhoff, 2001). Kansas egg dates range from May 1 to September 1.

Breeding Biology: As in other species, the male begins the breeding cycle by establishing a territory around a nest or nest site, which gradually shrinks to include only the nest itself and the immediate surroundings. Besides various threat postures, males also perform snap displays and a modified stretch display, called the "snap-hiss," accompanied by a raising of the ornamental crest plumes. These attract other birds, and eventually a female is allowed to enter the nest or approach the display site, after which the incipient pair begins mutual behavior such as nibbling and billing. Later the snap-hiss display serves as a greeting ceremony between the pair. The female completes the nest begun by the male, which may require up to a week. The first eggs are laid about three or four days after copulation, which may begin a day or two after the pair bond is formed. The incubation is by both sexes and begins with the first egg, so that hatching is staggered over several days. Until they fledge at about six weeks, and for a time afterward, the young continue to beg for food from their parents, following them to their foraging areas. Much of the foraging is done at night, which is the basis for the common name.

Comments: The night-herons are called thus because of their large eyes and associated abilities to forage late into the evening hours, when it is almost dark. This species is a fairly common breeder at Crescent Lake and Valentine refuges, and it is a startling sight, while walking through heavy marsh vegetation, to have one of these nearly invisible birds suddenly take flight from just a few feet away, National Breeding Bird Surveys between 1966 and 2009 indicate that the species underwent a statistically non-significant population decline (0.5 percent annually) during that period.

Suggested Reading:; Noble *et al.,* 1938; Palmer 1962; Hancock and Elliott, 1978; Hancock and Kushlan, 1984; B.O.N.A. 74; Kushlan and Hancock, 2005.

Yellow-crowned Night-Heron, *Nyctanassa violacea*

An uncommon spring and fall migrant and occasional summer visitor or nester in the eastern half of the region. Breeding is local in Kansas, rare in Nebraska and unproven in South Dakota.

Migration: Forty-three total spring sightings in Nebraska range from April 2 to June 10, with a median of May 6. Half of the records fall within the period April 29 to May 14. Twelve total fall Nebraska records are from August 1 to October 24, with a median of September 5.

Habitats: Like the black-crowned night heron, this species is found in diverse habitats ranging from saline to freshwater, and even breeds on rocky, nearly waterless islands. In our region it is usually associated with tree-lined river habitats.

Breeding Status: A local resident in southern and eastern Kansas, breeding at Cheyenne Bottoms Wildlife Area, and in various city parks, such as at Derby, Lawrence, Mulvane, Topeka and Wichita (Thompson *et al.,* 2011). There were no confirmed nestings during breeding bird atlas surveys in South Dakota (Peterson, 1995) or Nebraska (Mollhoff, 2001), but four in Kansas (Busby & Zimmerman, 2001). There is a single 1963 record of unsuccessful breeding in Sarpy County, and young fledglings have twice been seen near Lincoln (Sharpe *et al.,* 2001).

Nest Location: Nests in Oklahoma are usually in small, loose colonies separate from other heron species, in trees such as elms, ashes, oaks, box elders, and pecans. Nests there are 30-40 feet above ground level; in other areas the spread has been reported from no more than a foot above the ground to more than 50 feet. Old nests of the previous season are often used, and nests tend to be thick and well built, with materials added through the period of hatching.

Clutch Size and Incubation Period: In Kansas, the clutch is reported to be about four eggs, and 3-5 is the general range for the species. The eggs are pale bluish green with a smooth surface. The incubation period is 24–25 days. Sometimes double-brooded.

Time of Breeding: Kansas egg records are for the period May 16 to June 12. Eggs have been reported in Oklahoma as early as March 25, and broods have been seen as late as August 8. Two broods were raised by one pair in Norman, OK, in 1927, which fledged their first brood June 7 and the second two months later.

Breeding Biology: This species has been studied surprisingly little, but what is known suggests that it is very similar to the black-crowned night heron. The male evidently establishes a territory around a nest or nest site and advertises it with displays that probably include the stretch, accompanied by a loud whooping call. After pairs are formed, both sexes help complete the nest. Both sexes also incubate, and nest-relief ceremonies include billing, feather-nibbling, and plume erection or the stretch display. Both parents also feed the young, and their fledging period averages 37 days.

Comments: This beautiful night-heron mostly nests to the south of Nebraska, but occasionally finds its way into our state. Young birds resemble those of black-crowned night-herons, but are somewhat darker on the back and under wing-coverts. National Breeding Bird Surveys between 1966 and 2009 indicate that the species underwent a statistically non-significant population decline (0.7 percent) during that period.

Suggested Reading: Nice 1939; Palmer 1962; Hancock and Elliott, 1978; Hancock and Kushlan, 1984; B.O.N.A. 161; Kushlan and Hancock, 2005.

Family Threskiornithidae–Ibises

White-faced Ibis, *Plegadis chihi*

A rare to locally uncommon spring migrant and summer visitor throughout the three-state-region. Nebraska breeding has occurred in Clay County (in 1916), Garden County (1984, 1988, 1998) and Cherry County (Mollhoff, 2001). It has been observed at many times in Clay, Douglas-Sarpy, Garden, Lancaster, Phelps, Scotts Bluff, Sioux and York counties, and less often in Adams, Antelope, Brown, Cherry, Dakota, Dawes, Keith, Lincoln and Platte counties. Large numbers have been seen in the eastern Rainwater Basin during late summer.

Migration: Thirty-two total records range from April 9 to October 14. The largest number (14) of the sightings have occurred in May, followed by

April (nine), and there are two records each for June, August, September and October.

Habitats: Non-breeding birds may occur in almost any wet or moist habitat, including marshes, flooded fields, wet meadows, and other areas having shallow water for foraging. Nesting usually occurs in shallow marshes having extensive emergent vegetation.

Breeding Status: Breeding at Cheyenne Bottoms Wildlife Area in Barton County, Kansas, was first noticed in 1930, and this ibis has been a regular breeder since 1962. Breeding at Quivira N.W. R. began in 1981, and now is regular at both sites, with as many as 100 pairs reported at Quivira. Only six Nebraska nesting records were known as of 2003 (the first had bee found in 1915), but there have been several recent ones in the Nebraska Sandhills and Rainwater Basin. In 2010 there were 12 nests at Harvard W.P.A., Clay County (*Nebraska Bird Review* 78:90). Breeding in South Dakota was first reported in 1978 at Sand Lake N.W.R., Brown County, and continued for several years. Since then there have been nestings in Codington, Day and Kingsbury counties (Tallman, Swanson and Palmer, 2002). There was one confirmed nesting during breeding bird atlas surveys in South Dakota (Peterson, 1995), one in Nebraska (Mollhoff, 2001), and three in Kansas (Busby & Zimmerman, 2001).

Nest Location: The species breeds colonially. Nests are on the ground in dense vegetation and are constructed of dry emergent plants. When built in bushes or in trees surrounded by water, the nest may have a substantial platform of twigs, but more leafy materials are present than in heron nests.

Clutch Size and Incubation Period: From three to four eggs, dull blue with a smooth or finely pitted surface. The incubation period is 21 days. One brood per season.

Time of Breeding: Nests in South Dakota have been reported from June 4 to August 1 (Peterson, 1995). In Nebraska, nests with eggs have been seen as early as June 4 (Mollhoff, 2001). Kansas egg records are for the period June 1 to July 22.

Breeding Biology: Remarkably little is known of the biology of this species. Monogamous pair bonds are formed, and both sexes help construct the nest, which takes about two days. The incubation begins with the last egg. Both sexes also incubate, and during nest relief they do mutual billing and preening and utter guttural cooing notes. The adults continue to add material to the nest during incubation and the fledging period, about six weeks. The adults feed the young by regurgitation, with the young inserting their bills into that of the parent, or at times disgorge food into the nest to be picked up by the young. By the time the young are 6–7 weeks old they fly with their parents to the foraging grounds, returning with them at night for roosting. They are independent by eight weeks of age.

Comments: This somewhat exotic-looking bird has become fairly regular at Crescent Lake N.W.R.; I have seen groups of up to 14 birds there in recent years. National Breeding Bird Surveys between 1966 and 2009 indicate that the species underwent a statistically significant population increase (2.6 percent annually) during that period.

Suggested Reading: Palmer 1962; Burger and Miller 1977; B.O.N.A. 130; Johnsgard, 2012.

FALCONIFORMES – HAWKS, EAGLES & FALCONS

Family Accipitridae–Hawks & Eagles

Osprey. *Pandion haliaetus*

An uncommon to occasional spring and fall migrant throughout the region, probably most common eastwardly, where more large rivers and reservoirs exist.

Migration: The range of 102 initial spring sightings in Nebraska is from January 1 to May 25, with a median of April 21. Half of the records fall within the period April 12 to May 1. Twenty-one final spring sightings in Nebraska are from April 7 to May 27, with a median of May 5. Twenty-two initial fall sightings are from August 28 to November 30, with a median of September 15. Half of the records fall within the period September 14–24. Seventeen final fall sightings are from September 17 to December 26, with a median of October 9.

Habitats: While on migration this species occurs along rivers, lakes and reservoirs that support fishes and have fairly clear water for foraging. Habitats: Ospreys also occur along coastlines throughout much of the world.

Breeding Status: In South Dakota historic nesting occurred between the 1880's and 1900 in southeastern areas of the state. During the 1990's there were nesting records from Custer, Lawrence and Pennington counties (Tallman, Swanson and Palmer, 2002). There is a single old Nebraska breeding record for Rockport, a now-defunct town near the Douglas–Washington county line (Bruner, Wolcott and Swenk, 1904). Recent unsuccessful at-

tempts (starting in 2008) have been made near Winter's Lake in Scotts Bluff County, and (in 2009) at Lake Keystone/Ogallala in Keith County. A several-year hacking program in southeastern South Dakota along the Missouri River is likely to produce a regional nesting population. With that in mind, it is more than likely that ospreys will begin regular nesting in Nebraska and South Dakota again, considering the many new reservoirs that have been formed in recent decades. Hacking efforts have also been made in Kansas, where some nesting efforts have been seen, but so far there have been no known successful nestings in that state (Thompson *et al.*, 2011). There was one confirmed nesting during breeding bird atlas surveys in South Dakota (Peterson, 1995), and none in Nebraska (Mollhoff, 2001) or Kansas (Busby & Zimmerman, 2001).

Nest Location: In Minnesota, nests are 30–90 feet above the ground, usually at the tops of dead or partially dead lowland conifers, but sometimes on artificial structures such as power line poles. They consist mostly of dead sticks and branches. Unlike eagle nests in the region, they are usually rounded rather than cone-shaped and are generally smaller; the cup is typically lined with lichens.

Clutch Size and Incubation Period: From one to four eggs, usually three. The eggs are white with grayish and bright brown markings. They are laid at intervals of 1–3 days. The incubation begins with the first egg; its duration averages about 37 days. Single-brooded.

Time of Breeding: In Minnesota, egg records are from May 6–15, hatching occurs about June 15, and young are present in the nest until the middle of August. The unsuccessful Nebraska nesting attempt near Scotts Buff was seen by the author on May 17, 2008.

Breeding Biology: In northern parts of their range ospreys arrive in late April as the ice is melting from their nesting grounds, and males soon begin courtship flights. These swooping and soaring flights may serve to attract females but also continue for a time after pair bonds are established or reestablished. Nest-building or repair of the old nest starts very soon, the male bringing most of the larger sticks and the female bringing in the lining materials as well as doing the final shaping of the nest. From the time she arrives until the young are nearly fledged, the female catches few if any fish and thus relies on the male for virtually all her food. Mating occurs on the nest site or a nearby branch and continues during the egg-laying period. Both sexes incubate, but the female undertakes most of the responsibility and does all the nighttime incubation. The eggs hatch at intervals of up to five days, which results in considerable differences in the sizes of the young. For the first month of brooding the female rarely leaves the nest, and the male does all the hunting. As the young approach fledging at about 55 days of age, the female may also help in hunting. After fledging the young continue to use the nest for roosting and as a feeding platform, but they soon attempt to catch fish on their own. They do not mature sexually until they are three years old.

Comments: The birds are now regular spring and fall migrants at Lake McConaughy and other larger reservoirs in the state. Breeding Bird surveys between 1966 and 2009 indicate that the species had a statistically significant population increase (2.6 percent annually) during that period. The North American population (north of Mexico) has been estimated at

212,000 birds (Rich *et al.*, 2004). There is also a large Old World and Australian population; this species has one of the broadest breeding distributions of all living birds.

Suggested Reading: Dunstan 1973, Green 1976; Poole, 1989; Johnsgard, 1990; B.O.N.A. 683.

Bald Eagle. *Haliaeetus leucocephalus*

An uncommon spring and fall migrant, breeding summer resident, and local winter resident in the three-state region, especially along the major rivers and reservoir areas. During winter bald eagles appear at lakes, reservoirs or larger rivers throughout the region.

Migration: Sixty-five initial fall sightings range from September 16 to December 31, with a median of November 29. Half of the records fall within the period November 16 to December 16. Eighty-eight final spring sightings in Nebraska are from January 8 to May 12, with a median of March 19. Half of the records fall within the period March 17 to April 2. A state-level analysis of four decade-long periods of Christmas Bird Counts (1967-68 to 2006–7) extending from North Dakota to the Texas panhandle indicated a late-December population peak in Kansas (Johnsgard and Shane, 2009).

Habitats: Bald eagles in Nebraska utilize ice-free areas of larger tree-lined rivers and reservoirs during winter periods, such as at Lewis and Clark Lake, Lake McConaughy, Johnson Lake and Harlan County Reservoir. At Lake McConaughy, maximum numbers occur in late January and February, with migrant eagles leaving the state shortly after ice-breakup in spring. Perching is usually done in tall cottonwoods near water. Breeding is largely confined to forested regions in the vicinity of lakes or larger rivers that support a good supply of fish. Of 221 Nebraska nests observed between 1973 and 2009, 85 percent occurred along rivers, with the most seen along the Platte (42), Missouri (37) and Elkhorn (15) rivers. About 11 percent were located along lakes or reservoirs (Jorgensen *et al.*, 2010).

Breeding Status: There are recent nesting records for South Dakota, Nebraska and Kansas. In South Dakota, after an early extirpation of breeders in the 1880's, the birds began nesting again in 1992. Since then there have been breedings in Bennett, Bon Homme, Brown, Gregory, Meade, Roberts, Spink, Union and Yankton counties (Tallman, Swanson and Palmer, 2002). Although bald eagles once bred regularly in eastern Nebraska, the first known modern-era nesting attempt was in 1973 in Cedar County (*Nebraska Bird Review* 41:76). Since the early 1990's nesting efforts have occurred every year. By 1998 nesting efforts were known from Boyd, Douglas, Gage, Garden, Nemaha, Pawnee and Sherman counties. By 2005, over 40 nests were occupied, and by 2007 there were 55 active nests scattered around the state, mostly along the Platte, Missouri and Elkhorn rivers, and by 2009 nests had been reported from 62 of Nebraska's 93 counties (Jorgenson et al., 2010). In 2010 the total number of occupied nests had reached 90. Breeding in Kansas has been reported for 24 counties, mostly in the central and eastern parts of the state, but west as far as Hodgeman County (Tallman, Swanson and Palmer, 2011). There were two confirmed nestings during breeding bird atlas surveys in South Dakota (Peterson, 1995), one in Nebraska (Mollhoff, 2001), and three in Kansas (Busby & Zimmerman, 2001).

Nest Location: The recent nests in North Dakota and Nebraska have been in large cottonwoods. They are built of large branches picked up from the ground or broken off dead trees. The nest gradually increases in size with each year's use and is generally about 4-7 feet in diameter and up to ten feet thick in old nests. It is lined with aquatic vegetation such as cattails and bulrushes, or with other soft, leafy materials.

Clutch Size and Incubation Period: From one to three eggs, usually two. The eggs are dull white with a rough surface. The incubation period is 34–36 days. Single-brooded.

Time of Breeding: In Kansas, eggs are usually laid by late January, with young hatching in March through May, and fledging three months later, from May to July (Thompson et al., 2011).

Breeding Biology: After maturing and acquiring the adult plumage at four or five years of age, eagles pair monogamously and remain paired permanently. They perform aerial displays, one of which involves locking talons and tumbling downward through the sky for several hundred feet. In Minnesota these flights occur in March, or during the nest-building period. Copulation occurs at the same time, and egg-laying soon follows. Both sexes assist in incubation, and the young hatch at intervals of several days. The female and young are brought food by the male, which in Minnesota consists primarily of bullheads and suckers rather than important game fish. As the birds grow, both parents gather food for them, but rarely do more than two eaglets survive to fledging. This occurs at about 70 days of age, but the young birds follow their parents until they are evicted from the area by the adults. Of 440 active nests in Nebraska, 549 young fledged (1.48 per nest). At least 24 active nests were blown down by strong winds (Jorgensen et al., 2010).

Comments: Bald eagle winter populations have greatly increased in recent years; now an average of more than 1,000 birds winter within the state. Lake McConaughy is especially favored, but Johnson and Harlan County reservoirs, Calamus Reservoir, the J-2 hydroplant near Lexington, the central Platte River, and the Republican and Missouri rivers are also regularly used areas. As many as 500 birds have seen at a single location (Calamus Reservoir). Usually about 25–30 percent of these birds are immatures, suggesting that favorable reproduction is occurring. National Breeding Bird surveys between 1966 and 2009 indicate that the species underwent a statistically significant population increase (4.8 percent annually) during that period. The North American population (north of Mexico) has been estimated at 330,000 birds (Rich et al., 2004).

Suggested Reading: Dunstan et al., 1975; Sherrod, et al, 1976, Johnsgard, 1990; B.O.N.A. 506.

Northern Harrier. *Circus cyaneus*

A common migrant and permanent resident throughout southern parts of the three-state region, and a summer resident and breeder to the north. It is probably most common as a breeder in the glaciated prairie area of eastern South Dakota, in the Nebraska Sandhills and in the central mixed-grass plains of Kansas.

Migration: Uncommon to rare during winter in South Dakota, but a year-around resident in Kansas. Although in cold winters most birds may also leave Nebraska, in many areas and years the species can be regarded as a resident. Thirty-nine initial spring sightings in Nebraska range from January 1 to June 2, with a median of March 13. The wide spread of the records suggest it is a resident over much of the state. Thirty-six final fall Nebraska records are from September 14 to December 31, with a median of December 9. A state-level analysis of four decade-long periods of Christmas Bird Counts (1967-68 to 2006–7) extending from North Dakota to the Texas panhandle indicated a late-December population peak in Kansas (Johnsgard and Shane, 2009). Adult males are last to migrate south in the fall, and the first to return north in the spring.

Habitats: This species occurs in open habitats such as native grasslands, prairie marshes and wet meadows. Nesting is done in grassy or woody vegetation ranging from upland grasses and shrubs to emergent vegetation in water more than two feet deep.

Breeding Status: Breeds nearly through the region. There were 14 confirmed nestings during breeding bird atlas surveys in South Dakota (Peterson, 1995), ten in Nebraska (Mollhoff, 2001) and 18 in Kansas (Busby & Zimmerman, 2001). There are breeding records from more than 30 Kansas counties (Thompson et al., 2011).

Nest Location: Nests are in grassy vegetation, ranging from upland situations to wetland habitats including emergent plants such as cattails, bulrushes, and whitetop standing in water up to 2½ feet deep. North Dakota nest sites also include locations in shrubby willows along wet meadows or swamps and in patches of upland shrubs such as wolfberry, silver berry, and rose. The nest is constructed of sticks, twigs, and grasses and is up to 30 inches in diameter, without specific lining materials.

Clutch Size and Incubation Period: From 4–6 eggs, usually five. The eggs are white to pale bluish white, usually unmarked, but sometimes with pale brown spots. The incubation period is 24–30 days, usually beginning before the clutch is complete. Single-brooded, but renesting is frequent.

Time of Breeding: North Dakota egg dates range from April 26 to June 25, and nestlings have been recorded from June 15 to July 15. Nests in South Dakota have been reported from May 13 to July 19 (Peterson, 1995). One Nebraska egg record is from June 15 (Mollhoff, 2001). Kansas egg dates are from April 11 to July 6.

Breeding Biology: Males migrate separately from females and arrive on the nesting grounds first. They display aerially by performing a series of spectacular dives and swoops, especially in the presence of females. Later the pair may display in this way and also by locking talons in flight. The nest is constructed mainly by the female, though the male may help gather materials. Frequently the birds are semi-colonial, with up to six nests concentrated in a square mile. The eggs are laid at intervals of several days, and the female may begin to incubate at almost any time during the prolonged egg-laying period. Males feed their incubating mates, and on the basis of a group of six nests studied in Manitoba, sometimes provide food for two females. The young hatch at staggered intervals and while they are very small are brooded continuously by the female while the male brings in food. Later the female also hunts, but she usually receives by aerial transfer the food the male brings in. She is the only parent to feed the young directly. Where males are tending two nests the females must do more hunting by themselves, and starvation of young nestlings is frequent. The young fledge at about five weeks, males averaging a few days sooner than females.

Comments: Northern harriers are graceful predators that are usually seen sweeping low over marshes and fields, and showing white rump patches in both sexes. National Breeding Bird Surveys between 1966 and 2009 indicate that the species underwent a statistically significant population decline (0.9 percent annually) during that period. The North American population (north of Mexico) has been estimated at 209,000 birds (Rich *et al.,* 2004). There is also a large Old World population that is known in England as the hen harrier.

Suggested Reading: Watson 1977; Simmons, 1983; Hammerstrom, 1986; Johnsgard, 1990; B.O.N.A. 210; Johnsgard, 2012.

GRUIFORMES – CRANES, RAILS & GALLINULES

Family Rallidae–Rails, Coots and Gallinules

Black Rail. *Laterallus jamaicensis*

An extremely rare spring and fall migrant in Nebraska, and a rare breeding summer resident in Kansas. Not yet reported from South Dakota.

Migration: The few Nebraska records extend from April 22 to September 20. Sharpe, Silcock and Jorgensen (2001) listed two documented and five undocumented spring reports, from April 22 to May 31. There are 15 summer reports, seven of which are from Crescent Lake N.W.R., from May 18 to September 6. The latest report is for September 20. There are 14 county records from Kansas, ranging from April 21 to October 29. There are historic records from Nemaha County, one seen in the Omaha Market, and Cuming County (Bruner, Wolcott and Swenk, 1904). Recent records include one possibly seen at Lake 11 near Omaha, Douglas County, September 20, 1979, and one reported on May 13, 1979, from Phelps County (*Nebraska Bird Review* 47:67). Nebraska is slightly north of its known breeding range, but in addition to the several summer records at Crescent Lake N.W.R. (*Nebraska Bird Review* 63:73), a calling bird was reported from Verdigre, Knox County, May 25, 1986 (*Nebraska Bird Review* 71:138), and one was heard calling at Facus Springs, Morrill County, on July 9, 2001 (*Nebraska Bird Review* 69: 114).

Habitats: In the Great Plains this species occupies marshy meadows that are heavily overgrown with sedges and grasses. Like the yellow rail, it is much more likely to be heard than seen. Observations in Kansas indicate that the preferred nesting habitat consists of shallow marshy areas with stable water levels. Tiny marshes appear to be used often, and nests are constructed of grasses, rushes and sedges.

Breeding Status: The breeding range of this tiny and elusive species is most uncertain, but it is an uncommon local summer resident in Kansas. Records there include breedings from eight counties, and eggs have been reported from June 6 to July 6 (Thompson *et al.*, 2011). Thus far, only Kansas (Barton, Finney, Franklin, Meade and Riley counties) can be defi-

nitely considered breeding range for this region, on the basis of current evidence. There were no confirmed nestings during breeding bird atlas surveys
in South Dakota (Peterson, 1995), Nebraska (Mollhoff, 2001), or Kansas
(Busby & Zimmerman, 2001).

Nest Location: Nests are on damp ground, in dense grass or sedge vegetation, or above water on a mat of grasses. The nest is typically arched over
with interwoven grasses and has a lateral entrance. The surrounding vegetation is usually 18–24 inches high, and the deep nest cup sometimes contains a few feathers.

Clutch Size and Incubation Period: From 6–13 eggs, probably averaging
about eight. The eggs are creamy to buffy with large reddish spots at the
larger end. The eggs are laid daily, and the incubation period is from 17–21
days. Hatching reportedly is synchronous, and the precocial young are able
to leave the nest the day they hatch. Probably single-brooded.

Time of Breeding: The available Kansas egg records are for the period
June 6 to July 6.

Breeding Biology: Very little is known of the breeding biology of this species. The male's best-known call is a metallic *kik-kik-kikker* or *kik-kik-ker*,
while the female is said to use a more cuckoo-like *eroo* note in response to her
mate. Very little is known of the specific aspects of behavior associated with
nesting, but apparently they are much like those of the yellow rail. The fledging period is not certain but evidently is somewhere between 14 and 24 days.

Comments: This species is even less frequently seen than the yellow rail,
although some people have told stories of sitting still beside a dense marsh
and seeing a black rail suddenly appear and nearly walk across their feet!
Playback of recordings of black rail calls at Crescent Lake N.W.R suggests
that a breeding population may occur there, but this remains to be proven.

Suggested Reading: Bent 1926; Todd 1977; Burt, 1994; B.O.N.A. 123.

King Rail. *Rallus elegans*

A summer resident that is rare in eastern Kansas, very rare in eastern Nebraska, and accidental in South Dakota.

Migration: Nine total spring Nebraska records are from April 2 to June
9, with a mean of May 6. Five fall Nebraska records are from July 10 to September 11, with a mean of August 7. Sharpe, Silcock and Jorgensen (2001)
noted that of 48 reports for all seasons, more than half are from April 26 to
June 15, and usually arrival is in late April. Fall reports indicate that departure occurs in September, with a few records as late as mid-December.

Habitats: In our region the king rail is generally associated with freshwater marshes. In an Iowa study the birds were found on shallow marshes

up to four feet deep, with abundant shoreline and emergent vegetation of grasses and sedges. They are often associated with muskrats, whose runs open up the vegetation and provide passageways for the rails.

Breeding Status: Breeds from South Dakota southward through eastern Nebraska and eastern Kansas. Two South Dakota nesting records (in 1952 and 1974) exist for Moody and Bennett counties (Tallman, Swanson and Palmer, 2002). This large rail has been seen several times in the Lincoln area, but no definite recent nesting records have been established for Nebraska. An old nesting record does exist for Douglas County. This rail probably breeds locally and rarely east of a line from Jefferson to Knox counties. It possibly also nests in the Clear Creek marshes at the west end of Lake McConaughy. In Kansas, there are breeding records from 15 counties, and nonbreeding records from 27 other counties (Thompson *et al.,* 2011). There were no confirmed nestings during breeding bird atlas surveys in South Dakota (Peterson, 1995) or Nebraska (Mollhoff, 2001), but there are nine from Kansas (Busby and Zimmerman, 2001).

Nest Location: Nests are in rather dense emergent vegetation. In Iowa, six nests were in such vegetation, including four in lake sedges and two in river bulrushes, and all were in water 4–18 inches deep. The nests are basketlike structures of dead herbaceous vegetation, with an overhead canopy of emergent plants.

Clutch Size and Incubation Period: From 8–14 eggs, often 11. The eggs are pale buff with a few darker brown spots. The incubation period is 21–22 days, starting near the end of the clutch. Apparently single-brooded, except perhaps in the southern states. Renesting probably occurs frequently after nest loss.

Time of Breeding: Kansas egg records are from May 14 to June 22.

Breeding Biology: The onset of the breeding season in king rails is marked by the males establishing territories and beginning their low-pitched mating call, *chuck-chuck-chuck,* which attracts unmated females. Males evict other male rails, even of such small species as soras, from their territories. They also choose the nest site and do most of the nest-building. Usually several brood nests are also built and later are used for brooding the chicks. Both sexes incubate, with most of the young hatching simultaneously. The young grow rather slowly and remain close to their parents for more than a month. They do not fledge until they are 9–10 weeks old, and during this fledging period the adults molt and become flightless for a time.

Comments: National Breeding Bird Surveys between 1966 and 2009 indicate that the species underwent a statistically significant population decline (3.6 percent annually) during that period.

Suggested Reading: Tanner and Hendrickson 1956; Meanley 1969; Tacha and Braun, 1994; B.O.N.A. 3; Johnsgard, 2012.

Virginia Rail. *Rallus limicola*

An uncommon spring and fall migrant and summer resident almost throughout the region. It breeds widely in the region except for the drier western areas.

Migration: Thirty-six initial spring sightings in Nebraska are from February 14 to June 1, with a median of May 8. Half of the records fall within the period April 29–May 16. Thirteen final fall sightings are from July 21 to October 13, with a median of September 16. A state-level analysis of four decade-long periods of Christmas Bird Counts (1967-68 to 2006–7) extending from North Dakota to the Texas panhandle indicated a late-December population peak in Oklahoma (Johnsgard and Shane, 2009).

Habitats: Marshes with extensive stands of emergent vegetation such as taller grasses (cattails, phragmites), bulrushes, and sedges are the primary breeding habitat of this species. Habitat needs of Virginia rails and soras appear to be virtually identical. However, at least in Iowa, Virginia rails tend to nest in cattails and eat more insects and duckweeds, whereas soras favor whitetop or sedges for nesting and include a larger proportion of seeds in their diets. Nests are built over wet ground or in shallow water among emergent vegetation.

Breeding Status: Although rather elusive, this species is evidently a fairly common breeder in wetlands throughout most of the region. There were three confirmed nestings during breeding bird atlas surveys in South Dakota (Peterson, 1995), six in Nebraska (Mollhoff, 2001), and two in Kansas (Busby & Zimmerman, 2001). Virginia rails probably nest regularly in Nebraska, especially in the Sandhills, but actual nesting records seem to be limited to Arthur, Cherry, Garden, Holt, Keith, Lancaster, Lincoln and Sheridan counties. Until the modification of Nebraska's Kingsley Dam for hydropower generation in 1982, breeding probably occurred on Lake Ogallala/ Keystone, Keith County.

Nest Location: Nests are built over wet ground or shallow water in stands of emergent vegetation. When nests are built over water, the water is rarely more than ten inches deep. In Minnesota, all of 17 nests found in one study were in cattails, usually within 20-30 feet of open water or other vegetational edges. In Iowa, lake sedge was found to be the most important cover for 27 nests, and a Virginia study also indicated a preference for sedges and grasses over cattails for nesting. The nest is typically lined with fine grassy material and has an overhead canopy of live emergent plants.

Clutch Size and Incubation Period: From 6–13 eggs, usually about eight. The eggs are buffy to white with a few brown spots near the larger end. The incubation period lasts 17–20 days, with an average spread of 3.3 days between the hatching of the first and last egg. Probably single-brooded, but some renestings have been reported.

Time of Breeding: South Dakota egg dates are from May 24 to June 4, with juveniles or nestlings seen from May 17 to August 14. Nebraska egg records are from May 15 to June 6, and young as early as June 30 (Mollhoff, 2001). Kansas egg records are for the period May 1 to July 20, with most eggs laid in early June, and young have been seen from May 23 to August 11.

Breeding Biology: Shortly after returning to their breeding grounds, males establish territories, which they proclaim by uttering their distinctive *ticket, ticket* calls and maintain by evicting other male Virginia rails, though they reportedly tolerate sora rails. They probably construct their nests in a few days, but like other rails they may also build several "dummy nests" that are later used as brood nests. Males perform bill-nibbling and courtship feeding of their mates and perhaps do most of the nest-building as well. Eggs are laid approximately daily, and incubation (by both sexes) begins near the end of the clutch, resulting in a slight scattering of hatching periods. The young leave the original nest soon after hatching and can fly in 6–7 weeks. When they are about 60 days old the parents begin to peck at them and evict them from their territories.

Comments: Until Lake Ogallala was modified to allow for increased hydro-power capabilities for Kingsley Dam in the 1980s, a veritable chorus of Virginia and sora rails could regularly be heard on summer evenings. National Breeding Bird Surveys between 1966 and 2009 indicate that the species underwent a statistically non-significant population increase (0.3 percent annually) during that period.

Suggested Reading: Tanner and Hendrickson 1954; Kaufmann 1971; Tacha and Braun, 1994; B.O.N.A. 173; Johnsgard, 2012.

Sora. *Porzana carolina*

A common spring and fall migrant and locally common summer resident nearly throughout the region. It breeds very widely in the Great Plains excepting the southern and southwestern areas, and occurs throughout during Migration:

Migration: Of 108 initial spring Nebraska records, the range is from March 10 to June 3, and the median is May 6. Half of the records fall within the period April 30 to May 12. Twenty-five final fall sightings are from July 27 to November 27, with a median of September 30.

Habitats: Much like the Virginia rail, the sora prefers marshlands that have extensive stands of dense emergent vegetation, especially tall grasses and grass-like plants, and fresh to slightly saline waters. The birds feed mostly at the surface on plant seeds rather than probing for invertebrates as is typical of Virginia rails.

Breeding Status: Locally common in marshes over eastern portions of South Dakota and Nebraska. Local and uncommon in Kansas, where there are breeding records for six counties. There were four confirmed nestings during breeding bird atlas surveys in South Dakota (Peterson, 1995), one in Nebraska (Mollhoff, 2001) and one in Kansas (Busby & Zimmerman, 2001). Until the modification of Nebraska's Kingsley Dam for hydropower generation in 1982, breeding probably occurred on Lake Ogallala/Keystone, Keith County.

Nest Location: Where Virginia and sora rails nest in the same marshes, sora nests tend to be in deeper water, averaging from about nine to 12 inches in depth. The nest is elevated several inches above the water level and is often hidden in cattails, bulrushes, or sedges. It is basketlike, with a deep cup and sometimes a lateral runway to water.

Clutch Size and Incubation Period: From 6–13 eggs, often about ten. The eggs are a rich buffy color with some darker spotting and are darker overall than those of Virginia rails. They are laid daily, and incubation begins at varied stages of clutch completion. The incubation period averages about 19 days, but the spread of hatching is from 3–13 days, averaging about seven. Considered single-brooded, but there is some evidence of double-brooding.

Time of Breeding: Nests in South Dakota have been reported from May 29 to July 20 (Peterson, 1995). A Nebraska egg record is from May 29 (Mollhoff, 2001). Kansas egg records are for the period July 20 to August 21.

Breeding Biology: Territorial male soras are more aggressive than Virginia rails, evicting individuals of that species as well as of their own. The "whinney" is the male advertisement call and peaks at the time egg-laying gets under way. Nest-building is probably by both sexes, and several "dummy nests" are usually constructed near the primary nest. Both sexes incubate, and as the first chicks hatch they are tended by one parent while the other incubates the remaining eggs. Compared with Virginia rail young, soras are fed and brooded for a relatively short time, which perhaps facilitates second broods in some circumstances. The young attain their full juvenile plumage by six weeks and can fly when only about 36 days old. By this time in late July the adults have become flightless and are replacing their wing and tail feathers.

Comments: This species is the commonest of the rails in Nebraska, but nesting records are rather few and far between. During the data-gathering period in the 1980's for Nebraska's first breeding bird atlas, there was only a single confirmed nesting, in Lincoln County (Mollhoff, 2001). National

Breeding Bird Surveys between 1966 and 2009 indicate that the species underwent a statistically non-significant population decline (0.3 percent annually) during that period.

Suggested Reading: Pospichal and Marshall 1954; Tanner and Hendrickson 1956; Tacha and Braun, 1994; B.O.N.A. 250; Johnsgard, 2012.

Common Moorhen. *Gallinula galeata*

An occasional migrant in eastern Nebraska and a rare summer resident. It breeds regularly in Kansas, where it is a local and breeder. In South Dakota it is very rare, with a single nesting record.

Migration: Sixteen initial spring Nebraska records are from March 23 to June 1, with a median of May 11. Half of the records fall within the period May 1–29. Three final fall sightings are from July 26 to September 29, with a mean of August 22.

Habitats: The favored summer habitat of this species consists of freshwater ponds and marshes having an abundance of emergent vegetation. Nests are usually placed above water or on land surrounded by water. Unlike the purple gallinule, it does not need floating vegetation for nesting.

Breeding Status: Breeds occasionally in Kansas, and rarely in Nebraska, and at least once in South Dakota (at Sand lake National Wildlife Refuge). There are a few old and scattered Nebraska breeding records (Cherry, Lincoln and Douglas counties), and more recent (1984–1989) records in Lancaster and Fillmore counties (Mollhoff, 2001). There were no confirmed nestings during breeding bird atlas surveys in South Dakota (Peterson, 1995), three in Nebraska (Mollhoff, 2001) and two in Kansas (Busby & Zimmerman, 2001).

Nest Location: Nests are in water, suspended above water, or on land surrounded by water. Deepwater nests usually have a ramp up the side, whereas those in shallow water or on land do not. In Iowa, 17 of 19 nests were in cattails, the others in bulrushes. The nest is constructed of emergent and aquatic plants and has a well-developed cup of finer vegetation.

Clutch Size and Incubation Period: From 5–10 eggs (13 Iowa clutches averaged 7.1); in England first clutches average about six eggs, and renests or second clutches are somewhat smaller. The eggs are buffy with small brown dots or spots. The incubation period is 21–22 days, starting (in first clutches) with the next-to-last egg, or (in later clutches) midway through the laying period. A regular renester and sometimes double-brooded.

Time of Breeding: Iowa nests are initiated between mid-May and late June. Nebraska egg records are from June 23 to July 7, and young as early as July 8 (Mollhoff, 2001). Kansas egg records are for the period May 22 to July 10. In Oklahoma egg dates are from May 15 to July 18, and young have been seen from July 2 to August 8.

Breeding Biology: Common moorhens are highly territorial birds and in some areas maintain winter core-areas that later expand to become breeding territories. Within the territories the birds build three kinds of structures: display platforms, egg nests, and brood nests. Up to five temporary display platforms are built early in the breeding season, and one or two egg nests are constructed a week or two before egg-laying. The male gathers most of the nest materials, and the female incorporates them into the nest. The eggs are laid daily, and both sexes incubate. The young of the first brood hatch nearly synchronously and are fed by their parents within an hour after hatching. Up to five brood nests are built after the brood hatches. The young are tended by both parents for varying periods; in one case a pair began a new nest only 26 days after hatching their first brood. The chicks fledge at 60–65 days of age and tend to disperse soon afterward.

Comments: Common moorhens (previously called "common gallinules") are moderately common birds, but are nearly as elusive as rails in most locations. Yet, in England, where they are fully protected, they are as fearless as coots and can be easily seen in park lagoons, such as in the heart of London. National Breeding Bird Surveys between 1966 and 2009 indicate that the species underwent a statistically non-significant population decline (1.4 percent annually) during that period.

Suggested Reading: Frederickson 1971; Wood 1974; Tacha and Braun, 1994; B.O.N.A. 685; Johnsgard, 2012.

American Coot. *Fulica americana*

A common to abundant spring and fall migrant and summer resident throughout the region. Sometimes it over-winters where open water exists. It breeds and migrates throughout the Great Plains in suitable Habitats. It occurs throughout Nebraska, but is most abundant in the Sandhills marshes

Migration: Seventy-four initial spring sightings in Nebraska are from February 4 to June 7, with median of March 29. Half of the records fall within the period March 19–April 24. Eighty-two final fall Nebraska records are from July 25 to December 31, with a median of November 2. Half of the records fall within the period October 14 to November 21. A state-level analysis of four decade-long periods of Christmas Bird Counts (1967-68 to 2006–7) extending from North Dakota to the Texas panhandle indicated a late-December population peak in Oklahoma (Johnsgard and Shane, 2009).

Habitats: A wide variety of wetlands, ranging from small ponds or large lakes and reservoirs are used throughout the year, but those that are fairly shallow and rich in submerged aquatic plants are favored. Coots sometimes also forage in wet meadows and on grassy shorelines of lakes or ponds. Nesting usually occurs in emergent vegetation.

Breeding Status: Pandemic throughout region. There were 53 confirmed nestings during breeding bird atlas surveys in South Dakota (Peterson, 1995), 32 in Nebraska (Mollhoff, 2001) and seven in Kansas (Busby & Zimmerman, 2001).

Nest Location: In North Dakota, hardstem bulrush is the predominant species of emergent vegetation used for nesting cover. Cattails and other bulrush species are also frequently used, and in an Iowa study cattail cover accounted for more than 250 of 320 nests studied. Nests are built over water ranging from five to nearly 60 inches deep and are floating platforms anchored to the surrounding vegetation.

Clutch Size and Incubation Period: From 5–15 eggs (502 North Dakota nests averaged 8.8, and 281 Iowa nests averaged 9.0). The eggs are buffy, with small brown spots. The incubation period is 23–27 days, with onset of incubation ranging from the first egg to the last egg, and hatching of the young is usually staggered. Usually single-brooded, but renesting is frequent and double-brooding has been reported in Utah and California.

Time of Breeding: North Dakota egg dates are from April 29 to August 13, and young have been seen from May 22 to September 15. Nests in South Dakota have been reported from May 1 to July 27 (Peterson, 1995). Iowa nests are initiated between mid-May and late June. Nebraska egg records are from June 2 to July 11, and young as early as July 3 (Mollhoff, 2001). Kansas egg records are for the period May 11 to July 8.

Breeding Biology: Coots are monogamous, with potential lifelong pair bonds, and spend much of their time in advertising and defending territories. These are established soon after arrival on the breeding grounds, and although the male patrols the territory at first, later both members of the pair defend it. Pairs also construct display platforms for copulation and, as the egg-laying period approaches, construct one or more nests for their eggs, as well as brood nests later on. Both sexes participate in incubation, with the male most often incubating at night. Unlike gallinules, coots seem to have no specific nest-relief ceremony. Hatching is typically staggered over several days. Apparently the male takes the major responsibility for brooding the young birds, although the female may take the first-hatched chicks and leave the male to incubate and tend the later hatchlings. The young begin to beg shortly after hatching, and soon begin to follow the adults during their foraging. After a month or so they are nearly independent, but they beg occasionally almost to the time they fledge, at about 75 days of age. If the adults begin a second clutch they may expel the young of the first brood from the area while they are still fairly young.

Comments: Coots have a bad "image" problem; they often are accused of being stupid, and hunters scoff at them as game birds. In the 1950's the Fish and Wildlife Service tried to popularize their hunting by increasing the daily bag limit and calling them "white-billed ducks," but this did little or nothing to enhance their popularity. Yet, they are fascinating to watch as they establish and defend their territories, and tend to their rather odd-looking chicks. National Breeding Bird Surveys between 1966 and 2009 indicate that the species underwent a statistically non-significant population decline (0.4 percent annually) during that period.

Suggested Reading: Gullion 1954; Fredrickson 1970; Tacha and Braun, 1994; B.O.N.A. 697; Johnsgard, 2012.

Family Gruidae–Cranes

Sandhill Crane. *Grus canadensis*

An abundant spring migrant in the Platte Valley from Grand Island to Lewellen, uncommon to rare elsewhere, west at least to the Clear Creek area of Lake McConaughy. It is generally less abundant in the fall. Other than the Platte Valley, the only statistically significant area of concentration is near the Harlan County Reservoir, which the birds have increasingly used in early spring. Elsewhere in the Great Plains it is a regular but less common migrant.

Migration: Fifty-seven initial spring sightings in Nebraska are from January 8 to May 1, with a median of March 1. Half of the records fall within the period February 10–March 20. Thirty final spring sightings in Nebraska are from March 9 to June 1, with a median of April 7. In recent years, spring arrivals have becomes significantly earlier, with birds that wintered along the Texas coast arriving at the eastern end of the Central Platte (near Grand Island) first, followed sequentially by those wintering farther west (Johnsgard and Gil, 2011). The cranes roost along stretches of the river geographically congruent with their relative wintering region geography. The largest birds, from wintering areas in eastern Texas, roost near Grand Island, and are probably headed for breeding areas in Minnesota. Those using the western stretch near North Platte are from wintering areas in southeastern Arizona. These are the smallest of the birds using the river, and migrate

the farthest They fly as far as 4,000 miles from their wintering grounds to breeding areas in northern Siberia, up to nearly 1,000 miles west of the Bering Sea coast. Fifty-five initial fall Nebraska sightings are from September 2 to November 24, with a median of October 8. Half of the records fall within the period September 28 to October 22. Fifty-three final fall sightings are from October 1 to December 31, with a median of November 5. A state-level analysis of four decade-long periods of Christmas Bird Counts (1967-68 to 2006–7) extending from North Dakota to the Texas panhandle indicated a late-December population peak in northwestern Texas (Johnsgard and Shane, 2009). During the 2009–2010 Audubon Christmas Bird Counts there were 23,000 sandhill cranes at Salt Plains N.W.R., Oklahoma, suggesting that the Great Plains wintering population may be gradually moving northward as regional winters ameliorate. As a reflection of this trend, during the unusually warm fall and early winter of 2011-2012, several thousand cranes remained in the central Platte Valley well into February.

Habitats: Slowly flowing rivers, with relatively bare bars and islands for roosting, and adjacent wet meadows and croplands for foraging, are used by this species during migration: The central Platte Valley (between Lexington and Grand Island) is evidently the optimum spring habitat for this species in the entire Plains area, but smaller groups concentrate near North Platte (several thousand), the Clear Creek marshes of Garden County (up to 14,000) and North Platte National Wildlife Refuge (several thousand). Spring concentrations in the central Platte valley are unequaled anywhere in North America, usually peaking at nearly 500,000 in late March (Johnsgard and Gill, 2010). Cranes require extensive areas of minimal human disturbance for their nesting. They have large territories that vary with population density but often exceed 100 acres, usually consisting of wet meadows that provide water, sites for feeding, nesting, and roosting, and brood-rearing areas. The extensive prairie marshes that once offered these features are now mostly drained, and human disturbance at the remaining ones is too severe for cranes.

Breeding Status: There were no confirmed nestings during breeding bird atlas surveys in South Dakota (Peterson, 1995), Nebraska (Mollhoff, 2001) or Kansas (Busby & Zimmerman, 2001). Since 1996, breeding had occurred in Nebraska's eastern Rainwater Basin on at least nine occasions by 2003 (*Nebraska Bird Review* 67:48; 70: 122–127; 71:113). Pairs have also bred or attempted to breed at Facus Springs (Chet & Jane Fleisback W.M.A.), Morrill County, with four years of nestings through 2008; and a pair nested in Rock County in 2006 and 2008 (*Nebraska Bird Review* 76:101). Nesting has also recently occurred at Kiowa W.M.A., Scotts Bluff County (*Nebraska Bird Review* 78:47). The sandhill crane historically (1897, 1910) bred locally in South Dakota, but there are no definite modern South Dakota nesting records, although there was an unverified report of nesting in 2008 (Johnsgard, 2011). There are no historic or modern records of Kansas nesting.

Nest Location: Studies in Idaho indicate that nests are usually either in shallow water (averaging about eight inches deep) or on the shoreline fairly near water (averaging about 15 feet away). In decreasing order of usage, they nest in wet meadow-marsh edge areas near shore, islands, dry upland meadows, marsh area far from shore, and artificial dikes. Old nest sites are rarely used in following years, but the nest is often placed near the old site. Nests on dry land are small and simply constructed, whereas those on water are more bulky, and constructed of any vegetation easily available in the immediate vicinity.

Clutch Size and Incubation Period: Normally two eggs (rarely one or three). The eggs are olive with darker olive or brown spotting. The incubation period is 30 days, starting with the first egg. Single-brooded, but with renesting frequent if the clutch is lost early in the incubation period.

Time of Breeding: Iowa egg dates are from May 2 to May 27. In Nebraska, adults tending young have been seen in July and August.

Breeding Biology: Cranes are monogamous, probably pairing for life after reaching reproductive maturity at 3–4 years of age. Some cases of "divorce" have been found, often as a result of a female choosing a new mate following failure of an earlier nesting attempt, or selecting a male holding a better territory. Upon returning to their breeding areas, pairs establish territories as early as 2–4 weeks before nest-building gets under way, typically choosing the same territory year after year, but frequently moving the nest location within the territorial boundary from year to year. Nest-building is done by both sexes and may take from a day to a week or more. The eggs are laid at two–day intervals, and both sexes participate in incubation, with the female apparently always doing the nighttime incubation. The eggs typically hatch 24 hours apart, and the chicks begin to feed immediately. The first-thatched chick is often taken away from the nest by one adult, while the other remains to hatch the second chick. Perhaps because the young "colts" are very aggressive toward each other, they are often brooded separately, and frequently the younger often dies from harassment by the older sibling. Fledging occurs at 67–75 days of age, and the family soon migrates as a unit. In contrast to whooping cranes, families typically join to form large flocks during both spring and fall migrations.

Comments: Sandhill cranes are the perfect harbinger of spring in Nebraska, they arrive with the break in winter weather, and their departure coincides with the leafing out of our flowering trees and shrubs. National Breeding Bird Surveys between 1966 and 2009 indicate that the species underwent a statistically significant population increase (4.9 percent annually) during that period, which is surprising in view of the large numbers of lesser sandhills killed annually by hunters (about 30,000, including crippling losses), and the several thousands of greater sandhills that are also now shot annually by hunters. Constraints on reproduction in sandhill cranes result from their long periods to reproductive maturity, and the fact that usually only a single chick per pair is raised successfully, which means that annual "recruitment rates" to the population are typically only about 6–8 percent. Of these, at least four percent are annually lost to hunting, with this mortality factor especially affecting the young, inexperienced birds. In non-hunted populations many sandhill and whooping cranes reach at least 30 years of age, and this age potential is even true of a few birds in hunted populations.

Suggested Reading: Littlefield and Ryder 1968; Drewien 1973; Tacha and Braun, 1994; B.O.N.A. 31; Johnsgard and Gill, 2010; Johnsgard, 2011.

ORDER CHARADRIIFORMES - SHOREBIRDS

Family Charadriidae–Plovers

Snowy Plover. *Charadrius nivosus*

Rare but increasingly frequent spring and fall migrant and breeder in Nebraska, and a local nester in Kansas. Breeding in Nebraska started in 1998 near Ponca along the Missouri River. In 2003 two of three nesting pairs fledged young at Lake McConaughy, which represented the fourth and fifth nesting records for the state (*Nebraska Bird Review* 71:114). Nesting continued there until 2009, when flooding eliminated the lake's shoreline habitat (Brown, Dinsmore, & Jorgensen, 2012).

Migration: Six spring Nebraska records for this species range from April 6 to May 17, with a mean of April 28. Five fall Nebraska records are from August 7 to September 7, with a mean of August 21. Sharpe, Silcock and Jorgensen (2001) noted that of 26 spring reports, nearly all fit a late April to mid-May pattern. Four fall reports are from August 7 to September 2.

Habitats: Migrants are found on mudflats, alkaline flats, sandy shorelines, and in shallow ponds. The barren salt plains area of southern Kansas and Oklahoma represents prime breeding habitat for this arid-adapted species, and sandy riverbeds or barren shorelines of reservoirs are used secondarily.

Breeding Status: A local summer resident central to southwestern Kansas, largely limited to saline flats and sandy riverbeds. Kansas had breeding records from eight central and southwestern counties, with as many as 75 breeding pairs reported from Stafford County and 20 from Barton County (Thompson *et al.*, 2011). Nesting studies at Lake McConaughy in 2004 and 2005 located nine nests and ten additional broods. Six of the nine nests hatched, and 32 of 35 chicks that were monitored were known to fledge (*Nebraska Bird Review* 73: 154–156). Between 1996 and 2003 breeding has occurred in Nebraska's eastern Rainwater Basin on at least nine occasions (*Nebraska Bird Review* 67:48; 71:113). Pairs have also bred or attempted to breed at Facus Springs (Chet and Jane Fleisback W.M.A), Morrill County, over four years through 2008; and a pair nested in Rock County in 2006 and 2008 (*Nebraska Bird Review* 76:101). Nesting has also occurred at Kiowa Wildlife Management Area, Scotts Bluff County (*Nebraska Bird Review* 78:47). There were no confirmed nestings during breeding bird atlas surveys in South Dakota (Peterson, 1995), or in Nebraska (Mollhoff, 2001) but there were three in Kansas (Busby & Zimmerman, 2001).

Nest Location: Nests are on rock, gravel, or sandy substrates and consist of a slight hollow lined with bits of debris. Occasionally the nests are clustered in loose colonies, and the birds sometimes nest near tern colonies.

Clutch Size and Incubation Period: Usually three eggs, sometimes two. The eggs are sand-colored or buffy with small black spots or lines. The incubation lasts 23–29 days, averaging 26 days.

Time of Breeding: Kansas egg dates range from April 24 to July 26. Oklahoma egg records are from April 29 to July 11.

Breeding Biology: After arriving on their breeding areas and establishing territories, males begin to advertise with various calls and displays including "scraping," a ritualized nest-building behavior. One of the other male displays is a slow' 'butterfly flight" with a trilling call. Although the birds commonly breed around salt water, they can drink no more saline water than other shorebirds and must obtain liquid by eating insects or other succulent foods. Thermal extremes are also common in their often vegetation-less and highly reflective environment. Thus, during hot weather parental activity increases, the birds spending most of their time standing over the eggs rather than sitting on them. Both sexes incubate. The eggs are laid about three days apart, but hatching is synchronous. Both sexes also defend the eggs and young, performing effective "broken-wing" diversionary behavior. The young fledge at 25-30 days of age.

Comments: Like the piping plover, the pale gray back color of this species matches that of dry sand, and makes the birds almost impossible to see when they are sitting on their eggs. It has been suggested that the black upper breast markings of these birds are examples of disruptive coloration, actually making them harder to see.

Suggested Reading: Purdue 1976; Boyd 1972; Johnsgard, 1981, 2011; Hayman *et al.,* 1986; B.O.N.A. 154.

Piping Plover. *Charadrius melodus*

An occasional to rare spring and fall migrant, and a local rare summer resident in South Dakota and Nebraska. In Nebraska, there are numerous older nesting records for the Niobrara, North and South Platte, Loup and Missouri rivers, but most of the recent records are for the Platte (especially from Dawson County eastward), the lower Niobrara and upper

Missouri, Loup and Middle Loup rivers, and at Lake McConaughy, Keith County. In South Dakota most breedings occur along the Missouri River and its western tributaries, especially alone Lake Oahe, but there is rare breeding in the northeastern pat of the state. Classified as a nationally threatened species.

Migration: Sixty-one initial spring sightings in Nebraska are from March 27 to June 1, with a median of May 3. Half of the records fall within the period April 21 to May 12. Five final fall sightings are from July 27 to September 5, with a mean of August 19. Sharpe, Silcock and Jorgensen (2001) noted that most birds have departed by early September, although there are records as late as October 24. Sightings of color-banded birds have been made along the Gulf Coast, from southern Texas to western Florida (*Nebraska Bird Review* 78:30–34).

Habitats: Breeding birds are usually associated with sparsely vegetated shorelines of shallow lakes and impoundments, especially those having bare sand or salt-encrusted areas of gravel, sand or pebbly mud. In North Dakota, piping plovers are associated with sparsely vegetated shorelines of shallow lakes and impoundments, especially those that have salt-encrusted areas of gravel, sand, or pebbly mud. Sand dunes with little or no vegetation are also used for nesting.

Breeding Status: Breeds uncommonly and locally in southern South Dakota (Missouri Valley from Lake Oahe south to Union County), and rarely in the northeast (Day, Codington and Kingsbury counties). In Nebraska, the species breeds along the lower and central Platte River, Crescent Lake N.W.R., the Loup and Middle Loup rivers, the lower Niobrara River, and at Lake McConaughy, Keith County. Recent National Breeding Bird Surveys indicate that 250–280 pairs breed in Nebraska, most of them in the central Platte Valley, at Lake McConaughy, and on spoil piles associated with gravel operations. In recent years Lake McConaughy has had the second-largest nesting population of piping plovers anywhere; only South Dakota's Lake Oahe has greater numbers of nests. Between 2003 and 2005 from 117 to 202 nests were found at Lake McConaughy (*Nebraska Bird Review* 73:101), and in 2009 over 200 nests were present (Brown, Dinsmore, & Jorgensen, 2012). There were 18 confirmed nestings during breeding bird atlas surveys in South Dakota (Peterson, 1995), 21 in Nebraska (Mollhoff, 2001), and one in Kansas (Busby & Zimmerman, 2001).

Nest Location: Nests are simply hollows in the sand, sometimes lined with pebbles, or scrapes in gravel or pebbly mud.

Clutch Size and Incubation Period: From two to four eggs (typically four in first clutches, sometimes three in renesting efforts). The eggs are buffy with dark brown spots. The incubation ranges from 27 -29 days, starting with either the third or the last egg. Single-brooded, but renesting usually occurs if the clutch is lost in the first half of the breeding season.

Time of Breeding: North Dakota egg dates range from May 19 to July 5, and dependent young have been seen from June 26 to July 27. Nests in South Dakota have been reported from May 8 to June 24 (Peterson, 1995). Nebraska egg records are from May 27 to July 4 (Mollhoff, 2001). Kansas egg dates range from June 11 to July 17.

Breeding Biology: Piping plovers are monogamous, but mate-changing in successive breeding seasons is fairly frequent, even when the original mate is still available. The eggs are laid every other day, and incubation responsibilities are about equally divided by the two sexes. In most nests the eggs all hatch on the same day, and within 2-3 hours the young have dried off and are able to leave the nest. They are brooded by the adults until they are about 20 days old, and although they can run very well they tend to crouch and freeze when approached. Adults of both sexes feign injury when their brood is threatened. Until they fledge at 30–35 days of age, the young remain within 400–500 feet of the nest.

Comments: This is one of the state's breeding species that is considered threatened in Nebraska. Changes in annual river flows of the Missouri and Platte rivers have destroyed much of its historic breeding habitat, although sandpit operations have provided some new opportunities. The total northern Great Plains population comprised about 1,250 pairs in the mid-1990s, of which Nebraska's component represented about 20 percent. A similar-sized population breeds along the East Coast. Based on 2009 data, the Nebraska breeding population then consisted of about 140 pairs on the entire Platte River, about 100 pairs on the Missouri River, 25 pairs on the Niobrara River, and about ten pairs on the entire Loup River system. The three-year running average for breeding success by Nebraska plovers was 1.24 chicks fledged per nest. The Great Plains recovery plan calls for 465 pairs to be present and maintained for 15 years in Nebraska (Mary B. Brown, pers. comm.).

Suggested Reading: Wilcox 1959; Stout 1967; Johnsgard, 1981, 2012; Hayman *et al.,* 1986; B.O.N.A. 2; Brown, Jorgensen and Rehme. 2008; Brown, Dinsmore, & Jorgensen, 2012.

Killdeer. *Charadrius vociferus*

A common to abundant spring and fall migrant and summer resident throughout the region. Also breeds and migrates throughout the Great Plains.

Migration: The range of 86 initial spring sightings in Nebraska is from February 11 to May 27, and with a median of March 13. Half of the records

fall within the period March 8–19. The range of 110 final fall Nebraska records is from August 18 to December 31, with a median of October 19. Half
of the records fall within the period September 27 to November 10. A state-
level analysis of four decade-long periods of Christmas Bird Counts (1967-
68 to 2006–7) extending from North Dakota to the Texas panhandle indicated a late-December population peak in Oklahoma (Johnsgard and Shane,
2009).

Habitats: This highly adaptable species often occurs on open fields during migration, but typically breeds near wetlands where there is exposed
ground nearby. The birds seem to prefer gravely, stony or sandy areas for
nesting probably because they offer camouflage for the eggs and the incubating bird, but also nest in a wide variety of locations, sometimes even in
garden plots and building rooftops.

Breeding Status: A pandemic summer resident throughout the region, locally common around marshes, streams, and other wetlands. There were 90
confirmed nestings during breeding bird atlas surveys in South Dakota (Peterson, 1995), 157 in Nebraska (Mollhoff, 2001) and 403 in Kansas (Busby &
Zimmerman, 2001).

Nest Location: Nests are often some distance from water, in a surprising
variety of locations. Of 13 North Dakota nests, three were on garden plots,
two on bare fields, two on heavily grazed native prairie, two on exposed
sand or gravel, two on bare lake shorelines, and one each in a stubble field
and an abandoned farmyard.

Clutch Size and Incubation Period: Nearly always four eggs, rarely three
or five. The eggs are buffy with extensive black or dark brownish spotting
and blotching. The incubation period is 24-26 days, starting with the laying
of the last egg. Sometimes double-brooded, especially toward the southern
part of the range.

Time of Breeding: North Dakota egg dates range from April 18 to June
21, and dependent young have been seen from May 19 to July 25. Nests in
South Dakota have been reported from May 5 to June 24 (Peterson, 1995).
Nebraska egg records are from May 7 to July 6 (Mollhoff, 2001), In Kansas, egg dates are from March 10 to July 14, with a double peak of nest dates
suggesting double-brooding. Oklahoma egg dates are from March 30 to July
28, and dependent young have been seen from April 16 to September 19,
also indicating double-brooding.

Breeding Biology: Although some birds are paired at the time they arrive
on their nesting areas in southern Canada, most arrive unpaired. Males advertise their territories in a variety of ways, such as uttering the familiar *killdeer* calls while flying with slow, deep wing beats, and by sham-nesting or
"scraping" displays resembling nest-building behavior. Such scraping dis-

plays are performed not only by unmated males but also before copulation, during hostile encounters, and during actual nest construction. Once pair bonds are formed, the pair remains together and both sexes defend their territory, although they may do some foraging outside the defended area. Both sexes also incubate the eggs and care for their young, but males tend to be more aggressive toward humans, while females vigorously evict other killdeers from the nest vicinity. The familiar injury-feigning display, or "broken-wing act," is primarily directed toward potential mammalian predators; large grazing mammals such as horses and cattle are more likely to be threatened or even attacked. Evidently the male undertakes most of the brooding duties, which last about three weeks. Fledging occurs by the time the young are 40 days old.

Comments: This is the most widespread and common plover in North America, often nesting well away from water and close to human population centers. Its conspicuous "kill-deer" calls can be heard in Nebraska from mid-March onward, and its defensive "broken-wing" behavior is familiar to every rural schoolchild. Jorgensen (2004) observed peak numbers occurring in the eastern Rainwater Basin during the first week of April, with total birds seen annually during the five-year study period ranging from 777–1,130. National Breeding Bird Surveys between 1966 and 2009 indicate that the species underwent a statistically significant population decline (1.1 percent annually) during that period. Morrison *et al.* (2001) estimated the species' total population at one million birds.

Suggested Reading: Bunni 1959; Phillips 1972; Johnsgard, 1981; Hayman *et al.*, 1986; B.O.N.A. 517.

Family *Recurvirostridae - Stilts and Avocets*

Black-necked Stilt. *Himantopus mexicanus*

A local migrant and summer resident breeder in west-central Kansas, and the Nebraska Sandhills and Rainwater Basin. Rare elsewhere in the region, but seemingly slowly expanding its breeding range.

Migration: Eight Nebraska records extend from April 30 to August 8. Five of the records are for the month of May. Sharpe, Silcock and Jorgensen (2001) report spring records from April 17 on through summer, and fall records extending to September 21. Kansas records extend from March 21 to September 10.

Habitats: Generally associated with alkali ponds and marshes, often those also used by avocets. In inland sites this species breeds around shallow alkali ponds and lakes, but it also is found coastally around brackish and freshwater ponds, on rice plantations, and in other habitats.

Breeding Status: A local breeder in west-central Kansas, mainly at Cheyenne Bottoms Wildlife Area since 1974, and Quivira N.W.R. since 1976. There are nesting records for six Kansas counties (Thompson *et al.* 2011). Breeding in Nebraska began in the 1980's at Crescent Lake N.W.R., and has occurred over several years near Lakeside, Sheridan County since at least 1985, as well as in Dawes County in 1994, in Hall County in 1998, at Funk Lagoon (Phelps County) in 2003, and at Harvard Lagoon and North Lake Basin (Seward Co.) in 2005. Breeding in the eastern Rainwater Basin began in 2005 and had occurred four times by 2008 (*Nebraska Bird Review* 76:101). There were no confirmed nestings during breeding bird atlas surveys in South Dakota (Peterson, 1995), two in Nebraska (Mollhoff, 2001) and four in Kansas (Busby & Zimmerman, 2001).

Nest Location: Nests are in small colonies, usually of about 6–10 nests, often in grass hummocks and always close to foraging areas. The nest may even be surrounded by water, on a floating platform of sticks and vegetation. At times the eggs are laid in a simple scrape with no lining.

Clutch Size and Incubation Period: Normally four eggs, sometimes three, and rarely five. The eggs are buffy to sandy with blackish blotches. The incubation period is 25–26 days. Single-brooded.

Time of Breeding: Nests in South Dakota have been reported from May 19 to June 19 (Peterson, 1995). A Nebraska egg record is from June 4 (Mollhoff, 2001). In Kansas, egg dates are from May 21 to July 16.

Breeding Biology: Like avocets, stilts form pair bonds gradually and without associated elaborate displays, through the persistent association of a female with a particular male, in spite of initial aggressiveness by the male. Stilts defend territories on their breeding grounds better than avocets do and advertise them by aerial displays. Copulation in stilts is preceded by slight ritualized breast-preening by both sexes, apparently identical to that of avocets. Nest-building is probably done by both sexes, and materials are added to the nest through incubation. In periods of rising water the nest may be raised considerably by such added materials, and both sexes apparently share incubation about equally. The incubation begins when the last or penultimate egg is laid, and lasts 25–26 days. The eggs hatch relatively synchronously, and the young remain in the nest no more than 24 hours. They are probably brooded for at least a week, and are independent at about four weeks.

Comments: This elegant, long-legged shorebird has been slowly increasing in Nebraska, and elsewhere in the Great Plains. It seems to favor shallow, alkaline ponds such as those found in the western parts of Crescent Lake N.W.R., where cinnamon teal and Wilson's phalaropes also often congregate. National Breeding Bird Surveys between 1966 and 2009 indicate that the species underwent a statistically non-significant population increase (2.6 percent annually) during that period. The North American population has been estimated at 150,000 birds (Morrison *et al.*, 2001).

Suggested Reading: Stout 1967; Hamilton 1975; Johnsgard, 1981, 2012; Hayman *et al.*, 1986; B.O.N.A. 449; Jorgensen and Dunbar, 2005.

American Avocet. *Recurvirostra americana*

A regular breeder in the western half of the region, and migrants may appear throughout it. It is common in western Kansas, in central Nebraska, primarily in the Sandhills and Rainwater Basin, and in western and northeastern South Dakota.

Migration: Eighty-two initial spring sightings in Nebraska range from April 2 to June 7, with a median of April 28. Half of the records fall within the period April 20 to May 6. Thirty-eight final fall sightings are from July 25 to November 17, with a median of September 4. Half of the records fall within the period August 25 to September 2.

Habitats: In Nebraska, avocets are associated with shallow ponds or marshes with exposed and sparsely vegetated shorelines, often in association with strongly alkaline waters. In North Dakota, breeding is usually limited to areas of shallow water with exposed and sparsely vegetated shorelines, most often associated with alkaline to sub-saline wetlands. Among 253 pairs studied, many were found on strongly saline alkali ponds and lakes, and very few occurred on freshwater ponds and lakes. Nests are placed in exposed locations on mud flats, sand bars and islands, with little or no surrounding cover.

Breeding Status: A summer resident in the western portions of the region, extending locally eastwardly eastern South Dakota, central Nebraska, central Kansas (Finney, Barton, and Stafford counties). There were 13 confirmed nestings during breeding bird atlas surveys in South Dakota (Peterson, 1995), eight in Nebraska (Mollhoff, 2001) and eight in Kansas (Busby & Zimmerman, 2001).

Nest Location: Nests are found on mud flats, sandbars, and islands, often only slightly above the water surface and with little or no associated vegetation. The nest is a simple scrape, with a lining of materials found in the immediate vicinity, and is most extensively lined in areas subject to flooding. Nests are often in loose colonies near favored foraging areas.

Clutch Size and Incubation Period: From two to four eggs (17 North Dakota nests averaged 3.1), typically four in completed clutches. The eggs are
buffy to olive buff with many darker spots. The eggs are laid at approximately daily intervals, and the incubation period averages about 24 days but
varies from 22 to 29 days. Single-brooded, but replacement clutches have
been reported.

Time of Breeding: Nests in South Dakota have been reported from May
26 to June 19 (Peterson, 1995). Nebraska egg records are from June 2–28
(Mollhoff, 2001), In Kansas, egg dates are from May 2 to July 29. North Dakota egg dates range from May 12 to July 5, and dependent young have been
seen from June 9 to July 27. Oklahoma egg dates are from May 15 to June 21.

Breeding Biology: In Oregon, avocets arrive on their breeding areas 15–
20 days before egg-laying, to establish territories and perform precopula-

tory courtship. They apparently form pairs in late winter, without associated elaborate posturing. Copulation is preceded by a rather simple breast-preening ceremony that may be initiated by either bird. Pairs form close bonds and forage together as well as defend their territory as a unit. Both sexes develop incubation patches and begin to incubate their clutch as soon as it is completed. Early in incubation the male spends more time on the nest than the female, but the female is more attentive later on. Incubation requires 22–24 days. The eggs hatch over a one or two-day period, and the young soon become very active, feeding themselves from the outset. They fledge in 4–5 weeks and soon thereafter the families begin to form flocks.

Comments: The American avocet is one of the most beautiful of American shorebirds, and can be easily seen in the western parts of Crescent Lake National Wildlife Refuge. In the summer of 1995 over 140 avocets were present there at Smith Lake, which was then being drained temporarily. Jorgensen (2004) observed peak numbers occurring in the eastern Rainwater Basin during the third week of April, with total birds seen annually during the five-year study period ranging from 46–304. Three years of fall counts varied from 1–164 total birds. National Breeding Bird Surveys between 1966 and 2009 indicate that the species underwent a statistically non-significant population increase (0.1 percent annually) during that period. The North American population has been estimated at 450,000 birds (Morrison *et al.,* 2001).

Suggested Reading: Gibson 1971; Hamilton 1975; Johnsgard, 1981, 2012; Hayman *et al.,* 1986; B.O.N.A. 275.

Family Scolopacidae - Sandpipers and Phalaropes

Spotted Sandpiper. *Actitis macularius*

A common spring and fall migrant and summer resident throughout the region.

Migration: The range of 105 initial spring Nebraska records is from March 3 to June 5, with a median of May 4. Half of the records fall within the period April 26 to May 3. Sixty-two-final fall Nebraska records are from July 26 to October 26, with a median of September 9. Half of the records fall within the period from August 2 to September 22.

Habitats: Throughout its stay in Nebraska, this species is associated with wetlands having exposed or sparsely vegetation shorelines or islands, and ranging from fairly rapidly flowing streams to still-water habitats. The shore-line features are apparently more important than the characteristics of the water.

Breeding Status: A breeding summer resident that is locally common throughout the region. There were five confirmed nestings during breeding bird atlas surveys in South Dakota (Peterson, 1995), 12 in Nebraska (Moll-hoff, 2001) and eight in Kansas (Busby & Zimmerman, 2001).

Nest Location: Nests are on the ground in rather open terrain, often some distance from water. Cover above the nest varies from grasses 6–30 inches tall to weeds or bushes, and the nest itself is a slight depression lined with dried grasses.

Clutch Size and Incubation Period: Usually four eggs, sometimes three or five (15 North Dakota nests averaged 3.9). The eggs are buffy with heavy spotting of dark brown. The incubation period is 20–22 days, usually 21, and starts with the last egg laid. Some females are sequentially polyandrous and may lay several clutches.

Time of Breeding: Nests in South Dakota have been reported from May 3 to June 22 (Peterson, 1995). Nebraska egg records are from May 12 to June 13 (Mollhoff, 2001), In Kansas, egg dates are for June 11–27.

Breeding Biology: Male and female spotted sandpipers arrive on their breeding grounds at about the same time, and pair bonds are formed ex-tremely rapidly during a period of intense aggression, especially among fe-males, which are larger and more aggressive than males. Females establish territories, and pairs are formed by males entering such territories and be-ing either accepted or expelled by unmated females. When a male leaves the shoreline area and enters nesting cover with a female, a bond has been formed, and the female may lay her first egg within five days of the male's arrival. The eggs are laid at approximately daily intervals, and by the time she lays the third egg the female begins to show a resurgence of sexual ac-tivity, with increased singing and territoriality. Although some females re-main monogamous and assist with incubation, others allow their first mates to undertake incubation duties and accept a second mate. Successive mating with as many as four mates in a single season has been found, and typically the female helps incubate the final clutch. The young birds leave the nest as soon as their feathers dry and reportedly are able to fly as early as 13–16 days after hatching.

Comments: This common sandpiper can often be seen along most of Ne-braska's waterways, where its teeter-totter behavior and distinctive flight, with strongly down-curved wing actions, make it easily recognizable. It is also the only Nebraska shorebird with spotted underparts, at least in breed-ing plumage. Jorgensen (2004) observed peak numbers occurring in the eastern Rainwater Basin during the second week of May, with total birds seen annually during the five-year study period ranging from 28–63. National Breeding Bird Surveys between 1966 and 2009 indicate that the species underwent a statistically significant population decline (1.3 percent annually) during that period. Morrison *et al.* (2001) estimated the North American population at 150,000 birds.

Suggested Reading: Hays 1973; Oring and Knudson 1973; Johnsgard, 1981; Hayman *et al.,* 1986; B.O.N.A. 289.

Willet. *Tringa semipalmatus*

An uncommon to locally common spring and fall migrant throughout the region, and a locally common summer resident in the Nebraska Sandhills and most of South Dakota. It also breeds commonly in northeastern and north-central South Dakota. The Nebraska Sandhills apparently represent the species' southern limits of breeding in the Great Plains.

Migration: The range of 104 initial spring sightings in Nebraska is from March 18 to June 10, with a median of April 27. Half of the records fall within the period April 19–May 5. Sixteen final fall sights are from August 10 to November 9, with a median of August 24. Half of the records fall within the period August 19–September 1.

Habitats: A rather wide variety of wetland habitats are used by breeding birds, including streams, ponds, and marshes or shallow lakes, provided that prairie vegetation is located nearby. Less often hayfields or croplands may be used for nesting. In North Dakota, willets are found in fresh to highly saline wetlands, streams, and seasonal to semi-permanent ponds and lakes, but with highest densities on brackish or sub-saline lakes and semi-permanent ponds.

Breeding Status: Breeds locally in prairies and wetlands in the northern part of the region north of Kansas, including glaciated portions of South Dakota, and the Nebraska Sandhills. There were four confirmed nestings during breeding bird atlas surveys in South Dakota (Peterson, 1995), 19 in Nebraska (Mollhoff, 2001) and none in Kansas (Busby & Zimmerman, 2001).

Nest Location: Nests are in prairie vegetation, often 100 to 200 yards from the nearest water. Of 12 North Dakota nests, eight were in native prairie, three were in cropland fields, and one was in tame hayland. The nests are usually in thick grass, with the grass blades bent down to help provide a nest base and other grass added for lining. Some nests have also been found in almost wholly exposed locations.

Clutch Size and Incubation Period: Typically four eggs, rarely five (15 North Dakota nests all had four). The eggs are grayish to olive colored with darker brown spots and blotches. Single-brooded, but renesting has been reported. Incubation lasts from 24–26 days.

Time of Breeding: North Dakota egg dates are from May 10 to June 21, and young have been seen from June 11 to July 30. South Dakota egg dates are from May 29 to June 22, and young have been seen to July 22.

Breeding Biology: Willets arrive on their breeding grounds several weeks before egg-laying and the group includes paired birds as well as unpaired ones. Courtship is relatively social, and this flocking tendency conflicts with the territorial behavior of males, which tends to space the population. Aerial displays are common, consisting of uttering the distinctive *pill-willet* call while moving the wings through a narrow arc. Sparring fights on the ground between males are also common. Precopulatory display consists of standing behind the female and similarly vibrating the open wings, thus displaying the white areas on them. After pair bonds are formed the male follows the female about, often spreading his tail and exhibiting the white feathers, while the female apparently chooses the actual nest site. The nests are usually spaced 200 feet or more apart, and the eggs are laid at intervals of one to four days. Typically the female incubates by day, and the male at night. The eggs all hatch about the same time, and the young are highly precocial. Apparently the parents abandon their offspring before the latter have fledged, at about 28 days, and begin to leave the region.

Comments: It is always a surprise to students unfamiliar with willets when these rather dull-looking birds take flight and suddenly expose their stunning white wing markings; the willet's earlier generic name signifies "bearing a mirror." Jorgensen (2004) observed peak numbers occurring in the eastern Rainwater Basin during the fourth week of April, with total birds seen annually during the five-year study period ranging from 90–357. National Breeding Bird Surveys between 1966 and 2009 indicate that the species underwent a statistically non-significant population decline (0.6 percent annually) during that period. The North American population has been estimated at 250,000 birds (Morrison *et al.,* 2001).

Suggested Reading: Tomkins 1965; Stout 1967; Johnsgard, 1981; Hayman *et al.,* 1986; B.O.N.A. 579' Johnson & Igle, varied dates.

Upland Sandpiper. *Bartramia longicauda*

An uncommon spring and fall migrant and local summer resident in natural grasslands nearly throughout the region. It nests in suitable habitats almost throughout the Great Plains.

Migration: The range of 108 initial spring sightings in Nebraska is from March 9 to May 9, with a median of May 2. Half of the records fall within the period April 24–May 10. Seventy-five final fall sightings are from July 21 to October 28, with a median of August 20. Half of the records fall within the period August 10–26.

Habitats: During summer, this species occurs on native prairies, especially mixed-grass and tall grass, on wet meadows, hayfields, retired croplands, and to a limited extent, on fields planted to small grains. Throughout the area this species' abundance has declined as the extent of land in native prairies has decreased in recent decades.

Breeding Status: Breeding occurs almost throughout the region. There were 36 confirmed nestings during breeding bird atlas surveys in South Dakota (Peterson, 1995), 60 in Nebraska (Mollhoff, 2001) and 82 in Kansas (Busby & Zimmerman, 2001).

Nest Location: In North Dakota, all of 183 nests in one study were in grassland, mostly native prairie. The nest is simply a slight depression in the ground, usually well hidden in thick grass, with grasses arched overhead to provide protection. It is lined with dried grasses to form a rather deep cup.

Clutch Size and Incubation Period: Typically four eggs, rarely three or five (all of 189 North Dakota nests had four eggs). The eggs are creamy to pinkish buff with reddish brown spotting on the rounded end. The incubation period averages 21 days. Single-brooded, but renesting following early clutch failure is probable.

Time of Breeding: Nests in South Dakota have been reported from June 3 to July 22 (Peterson, 1995). A Nebraska egg (and newly hatched chicks) record is from June 21 (Mollhoff, 2001). Egg dates in Kansas are from April 21 to June 10, with a peak of egg-laying in early May.

Breeding Biology: In North Dakota, the first spring arrivals appear about two weeks before the start of nesting and are usually paired birds. Territorial birds perform a flight display consisting of circling with quivering wing beats while uttering a musical purring or chattering call and finally diving abruptly back to the earth. In North Dakota nesting begins almost simultaneously, and the eggs are laid at approximately daily intervals. Both sexes incubate, and adults typically feign injury when discovered on the nest. There is a fairly long interval between the first pipping and the hatching of the last egg, which may vary from less than 24 hours to about three days. The chicks are brooded by both parents, and by the time they are 30 days old they are full grown and presumably have fledged.

Comments: One of the most typical and beautiful of the Sandhills breeders, the upland sandpiper provides a definition of grace when it lands on a fence post and momentarily lifts both wings in a ballet-like movement, before inserting them in its flank feathers and coming to rest. Its territorial song-flights are equally memorable. Once called the "upland plover," it is indeed rather plover-like in having a short bill and upland habitat preferences. Jorgensen (2004) observed peak numbers occurring in the eastern Rainwater Basin during the first week of May, with total birds seen annually during the five-year study period ranging from 13–44. National Breeding Bird Surveys between 1966 and 2009 indicate that the species underwent a statistically significant population increase (0.6 percent annually) during that period, a surprising trend as compared with other grassland breeders, which are nearly all declining. The North American population has been estimated at 350,000 birds (Morrison *et al.*, 2001).

Suggested Reading: Stout 1967; Higgins and Kirsch 1975; Bowen, 1976; Ales, 1980; Johnsgard, 1981, 2001; Hayman *et al.*, 1986; B.O.N.A. 580; Johnson & Igle, varied dates.

Long-billed Curlew. *Numenius americanus*

A common migrant and summer resident in western Kansas, western Nebraska, and western South Dakota. particularly in the Nebraska Sandhills and High Plains regions.

Migration: Eighty-three initial spring sightings in Nebraska range from March 7 to June 7, with a median of April 11. Half of the records fall within the period April 5–21. Twenty-eight final fall sightings are from July 22 to September 21, with a median of August 18. Half of the records fall within the period August 5–September 1.

Habitats: In Nebraska, this species is associated with Sandhills grasslands, short-grass plains, and other grassy environments offering extensive foraging and nesting opportunities. The eastern breeding limits reach at least Garfield and Holt counties, perhaps the eastern edge of the Sandhills. In South Dakota the greatest number of nesting records came from the western fifth of the state (Peterson, 1995). In Kansas, the shortgrass plains of the southwest provide the best remaining habitat. Nests often occur in prairie vegetation on upland slopes that are close to moist meadows for foraging.

Breeding Status: Breeds in western South Dakota, western Nebraska (the Sandhills), and western Kansas (Stanton and Morton counties). There were 18 confirmed nestings during breeding bird atlas surveys in South Dakota (Peterson, 1995), 35 in Nebraska (Mollhoff, 2001) and two in Kansas (Busby & Zimmerman, 2001).

Nest Location: Favored nest sites are damp, grassy hollows in prairie vegetation or long slopes near lakes or streams. The nest is simply a slight hollow lined with a varying amount of grasses or weeds. At times the birds nest in loose colonies, and they frequently place their nests beside dried cow dung, presumably for better concealment. In the Nebraska Sandhills, the proximity of potential upland nesting areas to moist meadows for foraging was found to be the most important criterion for nest sites.

Clutch Size and Incubation Period: Usually four eggs, sometimes five, and rarely more in multiple clutches. The eggs are mostly olive buff with variable spotting of darker browns. The incubation period is 27–28 days. Single-brooded.

Time of Breeding: The probable breeding season in North Dakota is from late April to early August, with a peak from early May to early July. Nests in South Dakota have been reported from May 1 to June 1 (Peterson, 1995). Kansas egg records are for May and June.

Breeding Biology: In the Nebraska Sandhills, long-billed curlews arrive by early April, usually in flocks of fewer than 12 birds. The rest of the month is spent in prenesting activities, including establishing core areas and foraging areas. Core areas typically consist of rolling sands and are advertised by extended flight displays and calling above the ultimate nest site. Meadows adjacent to nesting locations are used for foraging and are advertised by similar flight displays. The foraging area is a part of the defended territory, and other curlews are forcibly excluded. Both sexes incubate, and both sexes care for the brood. Hatching in Nebraska occurs about the first week on June. The fledging period is 32–45 days. By early August, the adults and juveniles will have departed from the area.

Comments: It seems probable that the Nebraska Sandhills represent the last major breeding stronghold of this species in the Great Plains. The author has counted more than 50 curlews in a single wet meadow in Crescent Lake National Wildlife Refuge during July, at a time when flocks were starting to gather prior to fall migration; western Nebraska has perhaps the highest breeding densities in the Great Plains. During a study of the curlew in the Sandhills, two breeding curlews were fitted with radio-collars that allowed satellite continuous tracking. One of the two flew from Nebraska to northern Texas in less than 24 hours. The other later took a week to get from Nebraska to southern Texas. Both of them spent the winter on the Gulf Coast of northeastern Mexico (Gregory, 2010). National Breeding Bird Surveys between 1966 and 2009 indicate that the species underwent a statistically non-significant population increase (0.4 percent annually) during that period. The North American population has been estimated at 20,000 birds (Morrison *et al.,* 2001), but more recent estimates have been in the vicinity of 50,000 to 150,000 (Fellows and Jones. 2009).

Suggested Reading: Bicak 1977; Fitzner, 1978; Allen, 1980; Johnsgard, 1981, 2001, 2012; Hayman *et al.*, 1986; Redmond & Jenni, 1986; B.O.N.A. 628; Johnson & Igle, varied dates; Fellows and Jones, 2009.

Marbled Godwit. *Limosa fedoa*

An uncommon to locally common migrant in Kansas and Nebraska, and a regular breeder in South Dakota.

Migration: The range of 117 initial spring sightings in Nebraska is from April 5 to May 26, with a median of April 29. Half of the records fall within the period from April 22–May 10. Eleven final spring sightings in Nebraska are from April 19 to May 23, with a median of May 7. Eleven total fall Nebraska records are from July 20 to October 24, with a median of September 9.

Habitats: Godwits use a variety of wetland habitats for breeding in North Dakota, including intermittent streams, ponds, and lakes ranging from fresh to strongly saline. Semi-permanent ponds and lakes appear to be the preferred habitat, followed by seasonal ponds and lakes, then miscellaneous wetlands. Extensive mud flats, wet fields, sand bars and the shorelines of impoundments are commonly used by migrating birds.

Breeding Status: A summer resident and breeder over the glaciated portions of South Dakota. Formerly more widespread and previously believed to breed in northern Nebraska, but there are only a few Nebraska breeding records for Sheridan County (*American Birds* 44:1153; 45:1134; *Nebraska Bird Review* 73:102). There were six confirmed nestings during breeding bird atlas surveys in South Dakota (Peterson, 1995), but none in Nebraska or Kansas.

Nest Location: Nests are usually in native grassland vegetation, sometimes at considerable distances from water. Often the nests are in grassy cover only a few inches high and consist of a simple depression in the ground, lined with dead grasses. In higher grass cover, grasses interwoven to form a canopy overhead conceal the nest.

Clutch Size and Incubation Period: Typically four eggs, rarely three or five. The eggs are buffy to olive with dull brown spotting and blotching. The incubation period is about 21–23 days.

Time of Breeding: North Dakota egg dates range from April 17 to June 22, and dependent young have been seen from June 7 to July 18. Nests in South Dakota have been reported from May 14–27 (Peterson, 1995).

Breeding Biology: Remarkably little is known of the breeding biology of this species. Females are appreciably larger than males and have consider-

ably larger bills. Observations in North Dakota by T. Nowicki indicate that the male incubates during the day and the female at night. Incubating birds are surprisingly close-sitting, and have been known to allow themselves to be picked up from the nest. However, humans in the nesting area are sometimes attacked from the air by all the godwits in the vicinity, as many as 50 birds. The young are led to water after hatching and soon begin feeding with groups of adults. The fledging period has not been established, but in the arctic-nesting Hudsonian godwit it is about 30 days, and in the European species fledging occurs at 28–34 days.

Comments: The name "godwit" comes from a traditional English name, meaning "a good thing (to eat)" This is the largest of all godwits, and certainly one of the most attractive. Jorgensen (2004) observed peak numbers occurring in the eastern Rainwater Basin during the fourth week of April, with total birds seen annually during the five-year study period ranging from 13–22. National Breeding Bird Surveys between 1966 and 2009 indicate that the species underwent a statistically non-significant population decline (0.3 percent annually) during that period. The North American population has been estimated at 171,500 birds (Morrison *et al.,* 2001).

Suggested Reading: Bent 1907; Roberts 1932; Nowicki 1973; Johnsgard, 1981, 2001; Ryan, *et. al.* 1984; Hayman *et al.*, 1986; B.O.N.A. 492; Johnson & Igle, varied dates.

Wilson's Snipe. *Gallinago delicata*

A common spring and fall migrant, and a rare or localized summer resident in the entire region. Migrants occur throughout the Great Plains and stragglers uncommonly over-winter in the region.

Migration: Eighty-one initial spring sightings in Nebraska range from January 1 to May 29, with a median of April 13. Half of the records fall within the period April 4–21. Twenty-three final spring Nebraska records are from April 12 to May 28, with a median of April 29. Thirty-seven initial fall Nebraska records are from July 21 to December 21, with a median of September 18. Forty-two final fall Nebraska records are from July 27 to December 31, with a median of November 12. The data suggest that over-wintering in Nebraska is rather rare in this species. A state-level analysis of four decade-long periods of Christmas Bird Counts (1967-68 to 2006–7) extending from North Dakota to the Texas panhandle indicated a late-December population peak in Oklahoma (Johnsgard and Shane, 2009).

Habitats: Migrating birds are associated with marshes, sloughs and other wetlands that support areas of mudflats or mucky organic soil where foraging by probing is readily performed. In North America breeding by this species is primarily associated with peatland habitats such as bogs, fens, and swamps, which in our region are generally confined to Minnesota. Farther south, snipes also breed in marshy habitats along ponds, rivers, and brooks, where mucky organic soil and rather scanty vegetation are to be found. Marshes rich in shoreline and emergent vegetation and are preferred over more open ones.

Breeding Status: South Dakota records are scattered and include Bennett, Brookings. Custer, Day and Marshall counties. It is known to have bred in Cherry, Garden, Garfield, Howard, Lancaster and Rock counties of Nebraska. It is regular during summer at the Clear Creek marshes, Garden County, in the Pine Creek drainage north of Smith Lake in Sheridan County, and in many Sandhills marshes. There are no breeding records for Kansas (Thompson *et al.*, 2011). There were two confirmed nestings during breeding bird atlas surveys in South Dakota (Peterson, 1995), and five in Nebraska (Mollhoff, 2001).

Nest Location: Nests are usually in rather wet locations, usually on a hummock in cover provided by mosses, grasses, or heather. When nests are built in grasses, the previous year's growth is interwoven to form a canopy. A lining of fine, dry grasses is also typical.

Clutch Size and Incubation Period: Typically four eggs, rarely three or five. The eggs are light to dark brown with heavy spotting or blotching of darker brown. The incubation begins with the last egg and lasts 17–20 days, usually about 18. Single-brooded, but renesting in some areas is probable.

Time of Breeding: In North Dakota the breeding season lasts from early May to mid-July, with eggs seen from May 20 to June 27 and young reported from June 9 to July 15. Nests in South Dakota have been reported from May 13 to May 22 (Peterson, 1995). A Nebraska egg record is from May (Mollhoff, 2001),

Breeding Biology: During migration the males fly in advance of the females and arrive on their breeding grounds up to two weeks before them. Males immediately establish territories and begin advertising them with several displays, especially "winnowing," an aerial display in which the bird dives at a 45° angle with the tail fanned horizontally and the wings quivering. The vibration of the outer tail feathers produces the distinctive tremulous sound, with the wings used as "dampers" to prevent excessive vibration. After females arrive there is a good deal of chasing, and the female may mate with several males before forming an association with one, which happens when she selects the nest site and begins to lay. The pair bond lasts only until the chicks are hatched, and during incubation the male may also court other females. Only the female incubates, but the male returns to the nest at the time of hatching and collects the first active chicks, leaving the last two or three to be cared for by the female. The chicks grow rapidly and can flutter short distances at two weeks, but they cannot make protracted flights until they are about three weeks old. When about six weeks old they begin to gather with other young in "wisps" that may number in the hundreds and begin to migrate south before the adults.

Comments: Snipes are rarely seen until they suddenly take off in a low, twisting flight, usually uttering a raspy *scaip* note. The strange, "winnowing" noises made by vibrating the tail feathers of territorial birds flying high overhead can be heard in various parts of the state and probably indicate nesting, but very few actual nests or chicks have been found. Jorgensen (2004) observed peak numbers occurring in the eastern Rainwater Basin during the first week of April, with total birds seen annually during the five-year study period ranging from 6–147. National Breeding Bird Surveys between 1966 and 2009 indicate that the species underwent a statistically non-significant stable population (0.0 percent change annually) during that period. Morrison *et al.* (2001) estimated the North American population at about two million birds. Until recently, the North American population was considered conspecific with the European species (*G. gallinago*), and both were known in English as the "common snipe."

Suggested Reading: Tuck 1972; Johnsgard, 1981, 2012; Hayman *et al.*, 1986; B.O.N.A. 417; Tacha and Braun, 1994.

American Woodcock. *Scolopax minor*

An uncommon to occasional spring and fall migrant in eastern parts of the region, and an uncommon and local summer resident. Breeding is regular in eastern South Dakota west to the Missouri River.

Migration: Thirteen total spring sightings in Nebraska range from March 12 to June 1, with a median of April 10. Thirteen total fall sightings are from September 12 to November 14, with a median of October 15. A state-level analysis of four decade-long periods of Christmas Bird Counts (1967-68 to 2006–7) extending from North Dakota to the Texas panhandle indicated a late-December population peak in Oklahoma (Johnsgard and Shane, 2009).

Habitats: Woodcocks are generally confined to young forests with scattered openings on rather poorly drained land, especially soils supporting a large population of earthworms that can be readily obtained by probing. Nesting cover is usually of hardwood or mixed hardwood and conifer trees but may also be dominated by brushy growth.

Breeding Status: A local summer breeding resident in South Dakota (breeding records from Brookings, Clay, Day and Marshall counties (Tallman, Swanson and Palmer, 2002). Besides some possible early nestings, there is a Sarpy County record for 1972 and for Hamilton County in 1978 (*Nebraska Bird Review* 47:59). There are also records for Dodge, Lancaster, Merrick and Sarpy counties (Sharpe *et al,* 2001). During the period 1984–1989 there were confirmed nestings in Burt, Cedar, Holt, and Stanton, counties (Mollhoff, 2001). Displaying birds have been regularly observed west to the vicinity of Scotts Bluff (Kiowa W.M.A.). There are scattered nesting records for 15 Kansas counties (Thompson *et al.*, 2011), and there were four confirmed nestings during breeding bird atlas surveys in South Dakota (Peterson, 1995), and five in Nebraska (Mollhoff, 2001).

Nest Location: Nests are usually within 500 feet of a male's territory and are typically less than 50 yards from the edge of woody cover. Of more than 200 nests studied in Maine, nearly half were in mixed hardwoods and conifers, and most of the rest were in pure alder or other hardwood cover. The nest is usually at the base of a small tree or shrub and is simply a slight depression in the soil with little or no vegetation lining it.

Clutch Size and Incubation Period: Nearly always four eggs, rarely three. The eggs are pinkish buff to cinnamon with darker brown spotting. They are laid daily, incubation beginning with the last egg and lasting 20–21 days. Single-brooded.

Time of Breeding: Nests in South Dakota have been reported from April 21 to June 16 (Peterson, 1995). A Nebraska egg record is from March 20 (Mollhoff, 2001). In Kansas, egg dates are from April 7 to May 28, and young have been seen as early as April 4.

Breeding Biology: Shortly after returning to their breeding grounds, males begin their distinctive dawn and dusk display flights from territorial "singing grounds." These consist of a series of calls, a hovering flight with "twittering" wing noise, and a zigzag flight back to earth accompanied by a series of liquid trilling notes. The male attempts to copulate with any females that are attracted to such singing grounds, and it is probable that no pair bond is established. The female locates her nest in the general vicinity of the singing ground but is not protected by the male, and she does all the incubation and brooding alone. Female woodcocks are noted for being extremely "tight" sitters and if finally forced off the nest will perform strong injury-feigning displays. The young are soon led from the nest and begin to feed on earthworms as early as three days after hatching. They can fly short distances by the time they are three weeks old, and most broods probably break up 6–8 weeks after hatching.

Comments: Seemingly woodcocks have become more frequent nesters in Nebraska recently; at least there are now many more sightings of displaying territorial males, nests or chicks. The evening song-flight of males is an ethereal experience, and recently has been reported regularly from the hike-bike trail bridge near Fort Kearney State Park, as well as many locations near Lincoln and Omaha. Moist ground and a ready supply of earthworms are major habitat needs. National Breeding Bird Surveys between 1966 and 2009 indicate that the species underwent a statistically significant population decline (1.4 percent annually) during that period.

Suggested Reading: Sheldon 1967; Godfrey 1975; Johnsgard, 1981, 2012; Hayman *et al.*, 1986; B.O.N.A. 100; Tacha and Braun, 1994.

Wilson's Phalarope. *Phalaropus tricolor*

A common to abundant spring and fall migrant, and a common summer resident, from South Dakota to central Kansas. Breeding occurs over much of Nebraska, especially in the Sandhills and Rainwater Basin.

Migration: A range of 115 initial spring sightings in Nebraska is from April 6 to June 6, with a median of May 2. Half of the records fall within the period April 25 to May 10. Thirty-eight final fall sightings are from July 26 to October 20, with a median of September 8. Half of the records fall within the period August 19–September 12.

Habitats: Breeding occurs in wet meadows near aquatic habitats ranging from flooded ditches to ponds and marshes or shallow lakes, especially somewhat alkaline ones. The presence of wet meadows apparently is the major habitat criterion for this species, which is found near fresh to highly saline water and is associated with watery environments ranging from ditches or river edges to seasonal, semi-permanent, or permanent ponds and lakes. Migrants forage by swimming in open water, capturing surface invertebrates while swimming in tight circles, which produces a vortex that draws invertebrates up from below.

Breeding Status: A summer resident in suitable habitats over most of the region, especially the wetland areas of South Dakota, and the Nebraska

Sandhills. Local in central and western Kansas, with specific breeding records for Barton and Meade counties, and summer records for Finney, Kearny, and Seward counties (Thompson *et al.,* 2011). There were seven confirmed nestings during breeding bird atlas surveys in South Dakota (Peterson, 1995), seven in Nebraska (Mollhoff, 2001) and two in Kansas (Busby & Zimmerman, 2001).

Nest Location: Nests are well hidden in wet meadows and sometimes also occur in grassy swales or on hummocky areas of shallow marshy habitats. The nests are scrapes in the ground, lined with dead grass built up into a cup about two inches thick. When placed over water they may be built up to a level about six inches above the water.

Clutch Size and Incubation Period: Normally four eggs, occasionally three. The eggs are buffy with a varying amount of darker spotting. The incubation period is 16–22 days, probably averaging about 20 days. The incubation is normally by the male, but there is no proof that females regularly produce more than one clutch.

Time of Breeding: Egg dates in North Dakota range from May 26 into July 8, and dependent young have been seen from June 7 to July 17. Nests in South Dakota have been reported from May 17 to July 3 (Peterson, 1995). A Nebraska egg record is from June 4 (Mollhoff, 2001). Kansas egg dates are for May and June.

Breeding Biology: Although female phalaropes are appreciably larger and more brightly colored than males, recent studies have cast doubt on the idea that they are regularly polyandrous. Pair bonds apparently are formed after the birds arrive on the breeding areas, during a period of behavior that is intensely aggressive but little indicative of typical territoriality. The female probably makes the nest scrape after the pair is formed, but the male adds the nest lining. The eggs are laid about 48 hours apart, and presumably the female plays no further role in parental care. The male incubates for the 20–21-day period, and he leads his brood from the nest to foraging areas only a few hours after they hatch. The fledging period has not been reported, but in the closely related northern phalarope it is less than three weeks.

Comments: Wilson's phalaropes are interesting for many reasons, not the least of which is their "sex reversal " traits in which females are larger, more colorful, and transfer incubation and brood-rearing chores to the male. They breed commonly around Border Lake, a highly alkaline lake at the western edge of Crescent Lake N.W.R. Jorgensen (2004) found this species to be the second most abundant spring migrant shorebird in the eastern Rainwater Basin. He observed peak numbers occurring in the eastern Rainwater Basin during the second week of May, with total birds seen annually during the five-year study period ranging from 771–8,826. National

Breeding Bird Surveys between 1966 and 2009 indicate that the species underwent a statistically non-significant population decline (0.3 percent annually) during that period. The North American population has been estimated at 1.5 million birds (Morrison *et al.*, 2001).

Suggested Reading: Hohn 1967; Kangarise, 1979; Johnsgard, 1981, 2012; Bomberger, 1984; Hayman *et al.*, 1986; Murray, 1983; B.O.N.A. 83; Johnson & Igle, varied dates.

Family Laridae - Gulls and Terns

Franklin's Gull. *Leucophaeus pipixcan*

An abundant spring and fall migrant, and a common breeder in northeastern South Dakota. A very rare or accidental summer resident in Nebraska. Stragglers sometimes are present during the summer in the Sandhills area.

Migration: Eighty-nine initial spring sightings in Nebraska range from March 6 to June 8, with a median of April 10. Half of the records fall within the period March 27 to April 21. Fifty-eight final spring sightings in Nebraska are from April 2 to June 2, with a median of May 14. Fifty-two initial fall sightings are from July 20 to October 24, with a median of September 7. Fifty-eight final fall sightings are from August 17 to December 20, with a median of October 17. Half of the records are for the period October 3 to November 2.

Habitats: Migrants are often found on plowed fields, often closely following the moving plow in large flocks. Large, relatively permanent prairie marshes with extensive stands of semi open emergent cover are the primary breeding habitat in North Dakota, with more limited usage of shallow river impoundments.

Breeding Status: Breeds in scattered colonies in northeastern South Dakota, where nesting records include seven counties (Tallman, Swanson and Palmer 2001), with colonies of 15,000–25,000 pairs regular at Sand Lake N.W. R. during normal years (Peterson, 1995). There have been occasional records for Garden County, Nebraska (*Nebraska Bird Review* 34:63; 35:32). There is a single breeding record for Kansas, at Cheyenne Bottoms Wildlife Area., in 1993 (Thompson *et al*, 2011). There were two confirmed nestings during breeding bird atlas surveys in South Dakota (Peterson, 1995), none in Nebraska, and one in Kansas (Busby & Zimmerman, 2001).

Nest Location: Nests are usually in emergent vegetation such as cattails, bulrushes, phragmites, and whitetop, in water as deep as 4–5 feet, and frequently among those of nesting black-crowned night herons. Emergent stands that are not extremely dense and that are close to open water are preferred. The nest is a floating mass of dead vegetation, anchored to live plants and with a well-formed cup.

Clutch Size and Incubation Period: From two to four eggs, usually three The eggs are mostly brown to greenish brown with darker brown blotching or scrawling. The incubation period has been estimated at 18–20 days. Single-brooded.

Time of Nesting: North Dakota egg records extend from May 23 to June 26, and young have been seen as early as June 11. Fledged young have been reported as early as July 11. Nests in South Dakota have been reported from May 21 to July 8 (Peterson, 1995).

Breeding Biology: Franklin gulls nest in colonies, and after they arrive in their nesting grounds in spring they display on the previous year's nesting site, though they often shift to a new colony before the start of nest-building. Pairs apparently pick their nest sites on the basis of horizontal visibility and relative aggression. Where emergent vegetation is thick, reducing visibility, nests are closer together than where vegetation is less dense, and aggressive interaction between adjacent pairs is reduced. Both members of the pair assist in incubation, and they continue to add materials to the nest through the breeding period, presumably because of its floating nature. Unlike many gull species, Franklin gulls do not eat the eggs or young of their own species, but minks, great horned owls, and northern harriers are major predators of young gulls as well as of adults. After hatching, the young remain on their nesting platform until they are 25–30 days old, and they do not learn to distinguish their own parents from other adult gulls until they are more than two weeks old. Likewise, parents will accept alien chicks less than about two weeks old, both of which suggest a slower development of parental and offspring recognition than is typical of ground-nesting gulls. Fledging occurs at 28–33 days.

Comments: The Franklin's gull is a "seagull" with a breeding range that is a thousand miles from the sea, and is closely associated with prairie marshes for nesting. During migration it occurs in large flocks, and breeding also is performed in colonies. National Breeding Bird Surveys between 1966 and 2009 indicate that the species underwent a statistically significant population decline (4.0 percent annually) during that period. The North American population was estimated at about 500,000 birds in the 1990's (B.O.N.A. 116).

Suggested Reading: Burger 1974; B.O.N.A. 116; Howell and Dunn, 2007.

Ring-billed Gull. *Larus delawarensis*

A common to abundant spring and fall migrant throughout the region, with stragglers sometimes remaining through the summer months. Nesting occurs locally in South Dakota.

Migration: Eighty initial spring sightings in Nebraska range from January 3 to May 15, with a median of March 16. Half of the records fall within the period March 5–26. Fifty final spring sightings in Nebraska are from March 12 to June 7, with a median of May 12. Forty-eight initial fall sightings are from July 20 to November 15, with a median of September 12. Fifty-seven final fall sightings are from August 25 to December 21, with a median of November 28.

Habitats: A wide variety of lakes, reservoirs, rivers, marshes and other wetlands are used by migrants. Breeding occurs in colonies on isolated and sparsely vegetated islands of lakes and impoundments, the colonies varying in size from a few birds to more than 1,000 pairs.

Breeding Status: In South Dakota it has bred in Roberts County, and still breeds at Bitter Lake, Day County. There were four confirmed nestings during breeding bird atlas surveys in South Dakota (Peterson, 1995), but none in Nebraska or Kansas. South Dakota nesting records include three counties (Tallman, Swanson and Palmer 2001).

Nest Location: Nests are usually on gravel or in matted vegetation on a flat substrate and are simple scrapes lined with readily available sticks, weeds, or grasses. In dense colonies nests may be less than a yard apart, and they are typically closer together than would be expected from random nest selection.

Clutch Size and Incubation Period: Normally three eggs, sometimes two or four. The eggs are buffy to whitish with darker brown markings. The incubation period averages 25 days. Single-brooded.

Time of Breeding: Egg dates in North Dakota extend from May 17 to June 27, and flightless young have been seen from June 8 to July 28. Fledged young have been seen as early as mid-July. Nests in South Dakota have been reported from May 8 to July 5 (Peterson, 1995).

Breeding Biology: Ring-billed gulls arrive on their nesting grounds well before the nesting season and establish nesting territories as early as possible, at times occupying exactly the same territory as in the previous year. Such behavior probably helps to maintain pair bonds and also results in birds returning to areas where successful breeding has previously occurred. As in other gulls, most pair-forming behavior consists of hostile postures and calls associated with territoriality. The eggs are laid at two-day intervals and, as in the California gull; egg-laying is highly synchronized within colonies. The incubation is by both sexes, and apparently a major source of egg mortality comes from chilling as a result of disturbance to the colony. Once the eggs hatch, most chick mortality evidently comes from pecking by neighboring adults when a chick wanders too far from its parents. The young birds fledge in an average of 37 days.

Comments: This is the most common white-headed gull in Nebraska. National Breeding Bird Surveys between 1966 and 2009 indicate that the species underwent a statistically significant population increase (13 percent annually) during that period. Its total world population has been estimated at 3–4 million birds, about 70 percent of which nest in Canada (Olsen and Larson, 2004; Alderfer, 2006).

Suggested Reading: Tinbergen 1959; Vermeer 1970; B.O.N.A. 33; Howell and Dunn, 2007.

California Gull. *Larus californicus*

An uncommon to rare migrant in the region, breeding locally in South Dakota.

Migration: Seven spring Nebraska records are from March 19 to April 26, with a mean of March 28. There is at least one June record. Ten late summer and fall Nebraska records are from July 18 to November 10. Nine winter records extend from December 13 to February 15. Sharpe, Silcock and Jorgensen (2001) reported spring records extending from March 19 to May 30, and fall records from July 18 to December 30. There are also some June records. Numbers gradually increase at Lake McConaughy during late summer and fall, and then decrease in early to mid-winter, with some birds persisting through winter.

Habitats: Lakes, large marshes and similar habitats are used by migrants. In North Dakota, this species nests in much the same habitats as does the ring-billed gull---barren islands on brackish or alkaline lakes---and the two species sometimes nest in mixed colonies.

Breeding Status: Limited as a breeder to northeastern South Dakota, where it has bred in recent years in at least Butte, Day, Marshall, McPherson and Perkins counties (Tallman, Swanson and Palmer, 2002). There was one confirmed nesting during breeding bird atlas surveys in South Dakota (Peterson, 1995), but none in Nebraska or Kansas.

Nest Location: Nesting is colonial; individual nests consist of scrapes lined with grasses, sticks, or weeds. In one North Dakota colony, the California gulls nested in the more central and elevated parts of an island, while the ring-billed gulls occupied the area near the water's edge. Likewise in Alberta, California gulls tend to nest on elevated and boulder-strewn areas, while ring-billed gulls occupy more level terrain. California gulls also tend to space their nests almost randomly, whereas ring-billed gulls show a tendency to aggregate.

Clutch Size and Incubation Period: Usually three eggs, but 2–5 have been reported. The eggs are white to buffy with darker spotting. The incubation period is 23–27 days. Single-brooded.

Time of Breeding: North Dakota egg dates are from May 3 to June 24, and flightless young have been seen from June 8 to July 28. Nests in South Dakota have been reported from May 28 to July 3 (Peterson, 1995).

Breeding Biology: California gulls arrive from their wintering grounds along the Pacific coast some weeks before the onset of nesting. Territorial establishment and courtship activities begin as soon as they arrive, even if the nesting areas are still covered by snow. Eggs are laid at an average interval of two days, so that most clutches are completed in 4–5 days. Egg-laying within colonies is highly synchronized, and in a sample of 100 nests nearly all the eggs were laid within two weeks. The incubation is performed by both sexes and averages about 26 days, with a range of 23–28. Although these gulls are serious egg predators for other species, relatively few eggs are eaten or disappear within the nesting colony, and hatching success is often high. The chicks are relatively precocial, and though they are usually raised in the close vicinity of their nest they are also well able to run and elude danger from an early age. They fledge at from 36–44 days of age, averaging 40 days.

Comments: California gulls closely resemble herring gulls, but have yellowish green or grayish green legs and are somewhat smaller. When seen beside herring gulls, their darker upperparts are also apparent. National Breeding Bird Surveys between 1966 and 2009 indicate that the species underwent a statistically non-significant population decline (0.5 percent annually) during that period. Its total North American population has been estimated at 50,000–100,000 birds (Olsen and Larson, 2004; Alderfer, 2006).

Suggested Reading: Vermeer 1970; Baird 1976; B.O.N.A. 259; Howell and Dunn, 2007.

Least Tern. *Sternula antillarum*

An uncommon spring and fall migrant in the region, and a local breeder along rivers and reservoirs in eastern and central parts of Nebraska, South Dakota, and Kansas. The local breeding population is of a federally endangered subspecies (*S. a. athalassos*).

Migration: Eighty-seven initial spring sightings in Nebraska range from March 8 to June 10, with a median of May 23. Half of the records fall within the period May 16–30. Twenty-six final fall sightings are from July 20 to October 6, with a median of August 14.

Habitats: Associated with rivers, lakes and impoundments on migration; nesting is mostly on river sand bars or islands, but sometimes also on barren shorelines of large impoundments, gravel beaches, or even newly cleared land. Nesting is typically done in colonies, on a sand or gravel substrate. Many birds are now nesting at sand and gravel mining operations, lakeshore housing developments and dredging operations as their natural nesting habitat (midstream river sandbars) is disappearing. In Oklahoma, nesting also occurs on salt plains, in habitats similar to those used by snowy plovers.

Breeding Status: Breeds locally and irregularly in the Missouri Valley of South Dakota, and the Missouri, Platte and Niobrara valleys of Nebraska. In recent years most Nebraska nesting has occurred along the Platte and Missouri rivers, with the Platte population extending west from its mouth to Lake McConaughy and the North and South Platte rivers in Keith County. The Missouri River population extends from Boyd County downstream to Dixon County, and west along the Cheyenne River. There have also been breeding birds along the Elkhorn River from Madison to Cuming County, and along the Loup River from Sherman County east to its Platte River confluence, as well as on the North Loup in Valley County (Panella, 2010; Mary B. Brown, pers. comm.). Nesting at Lake McConaughy occurs at Martin and Arthur bays, as well as along the south shore (Brown, Dinsmore, & Jorgensen, 2011). There were 22 confirmed nestings during breeding bird atlas surveys in South Dakota (Peterson, 1995), 24 in Nebraska (Mollhoff, 2001) and four in Kansas.

Nest Location: Nests are usually in colonies and consist of a simple scrape in sand or gravel, with little or no lining. Solitary nesting is frequent in the Great Plains.

Clutch Size and Incubation Period: From two to four eggs, but typically two. The eggs are pale buffy with darker brown spotting. The incubation period is 20–21 days. Apparently single-brooded, but renesting has been reported in coastal areas.

Time of Breeding: A few North Dakota egg records are from July 2 to July 21. Nearly all nests in South Dakota have been reported from June to July 8 (Peterson, 1995). Nebraska egg records are from June 3 to Jul 4 (Mollhoff, 2001). Kansas egg dates are from May 21 to June 30, with the modal date of egg-laying being June 5.

Breeding Biology: In the Mississippi and Missouri valleys, least terns usually arrive in May, sometimes before sandbars suitable for nesting have been exposed by declining river levels. The exposure of these bars sets nesting in motion and thus synchronizes the breeding cycles of each nesting colony. During courtship a bird may make aerial glides while carrying fish, then alight and offer the fish to another bird. Sex recognition may be achieved in this way, since if a male is offered a fish it responds by attacking. Incipient nest-building by the male may stimulate the female to begin the actual nest, which is a simple scrape in the sand. Nest sites are usually widely spaced, lessening antagonism between nesting pairs. The eggs are laid on consecutive days or at two-day intervals, and incubation probably begins with the first egg. At first the female incubates alone, but gradually the male assumes part of this duty. The eggs typically hatch on consecutive days, and the female does most of the brooding. Within a day the chick and parent have learned to recognize each other, and thus the parents feed no young other than their own. Within two days after hatching the young begin to wander away from the nest and usually do not return. They fledge on about the 20th day after hatching, and the colony is gradually deserted.

Comments: National Breeding Bird Surveys between 1966 and 2009 indicate that the species had a statistically non-significant population decline (2.3 percent annually) in that period. During drought conditions along the Platte River in central Nebraska, least terns and piping plovers moved their nests from riverine sites to sandpits, without apparent reducing their nesting success (*Nebraska Bird Review* 73: 71–78). The fledging rate (number of independent tern chicks raised per nest) in the lower Platte River valley ranged from 0.49 to 1.15 between 2008 and 2010. In 2006 the Platte Valley accounted for 4.4 percent of the Interior race's population, and 7.4 percent of its total colonies (C. A. Lott, *via* M. B. Brown). The total North American population was estimated at more than 50,000 birds in the 1990's (B.O.N.A. 290).

Suggested Reading: Hardy 1957; Tompkins 1959; Kirsch, 1992; B.O.N.A. 290; Johnsgard, 2012.

Caspian Tern. *Hydroprogne caspia*

An uncommon spring and fall migrant in eastern parts of the region, especially along the Missouri River. There are two breeding records for South Dakota, in Day and Dewey counties (Tallman, Swanson and Palmer, 2002).

Migration: Twenty-seven total spring sightings in Nebraska are from March 23 to May 28, with a median of May 10. Half of the records fall within the period May 3–17. Twenty-four fall Nebraska records are from July 20 to October 14, with a median of September 19. Half of the records fall within the period September 4–25. Sharpe, Silcock and Jorgensen (2001) noted that spring reports extend from April 22 to June 4, and fall reports from July 18 to October 9. There are also many reports of summering non-breeders in the region, and some wintering occurs north to Kansas.

Habitats: Larger rivers, deep marshes, lakes and reservoirs are used by migrants and summering non-breeders. A few non-breeders often summer in the Lake McConaughy region. Regularly reported in Nebraska as far west as Garden, Keith, Lincoln and Sheridan counties, and in Kansas to Finney and Morton counties. Most breeding is near the coastline, usually on sandy or stony beaches, but breeding also occurs on offshore islands and sometimes on the shorelines of large inland lakes.

Nest Location: Nests are on the ground, on sand, shingle, or shell beaches. The nest is a shallow hollow, usually unlined but sometimes has a slight accumulation of plant debris. Nesting usually occurs in colonies, but at times single pairs are found in the vicinity of other tern species.

Clutch Size and Incubation Period: Usually two eggs, sometimes three, rarely one. The eggs are creamy to creamy buff with dark specks, spots, and blotches that tend to be small and rather evenly distributed. The incubation period is about 26–28 days and begins with the first egg. Single-brooded, but probable replacement clutches have been noted.

Time of Breeding: In the case of the North Dakota nesting, the young were still mostly downy when seen on June 28 but had evidently fledged by July 12. The only definite breeding records are from South Dakota, where a bird on its nest was seen on June 22, and July 16, when an adult and chick were observed.

Breeding Biology: Nesting colonies of this species on islands in northern Lake Michigan and Lake Huron tend to be rather large, averaging about 150 pairs and sometimes reaching as many as 500 pairs. The colonies are also densely packed, with territories averaging less than two square yards, and the centers of adjacent nests may be as little as 21 inches apart. Incubation is by both sexes, and both parents tend the young. The fledging period has been estimated to be as little as 4 weeks by some authorities and as long as 6-8 weeks by others. After fledging, the juveniles rapidly disperse and gradually work their way to coastal wintering grounds. Immature birds are very sedentary and may spend a full year on the wintering grounds. The birds become adult toward the end of their third year of life.

Comments: National Breeding Bird Surveys between 1966 and 2009 indicate that the species underwent a statistically significant population increase (2.6 percent annually) during that period. The North American population was estimated at about 33,000–35,000 pairs in the 1990's (B.O.N.A. 403).

Suggested Reading: Ludwig, 1965; B.O.N.A. 403; Howell and Dunn, 2007.

Black Tern. *Chlidonias niger*

An abundant spring and fall migrant throughout the region, and a locally common summer resident, primarily in the Sandhills but locally elsewhere as well. Breeds from central and eastern South Dakota south to central Kansas, and a migrant throughout the entire region.

Migration: The range of 130 initial spring sightings in Nebraska is from April 9 to June 5, with a median of May 12. Half of the records fall within the period May 6–18. Sixty-six final fall sightings are from July 21 to October 5, with a median of September 2. Half of the records fall within the period August 19 to September 11. The more precise nature of this species' migration, as compared with the other terns and gulls, is no doubt a reflection of its insectivorous diet.

Habitats: Migrants are found over a variety of aquatic habitats, and sometimes also forage well away from water over adjoining grasslands. Favored breeding habitats are small to large marsh areas containing both extensive stands of emergent vegetation and areas of open water. Unlike common and Forster terns, this species feeds predominantly on insects and thus does not compete strongly with fish-eating species.

Breeding Status: Breeds locally over eastern South Dakota, the Nebraska Sandhills and other permanent wetlands of Nebraska. There is only one definite nesting locality for Kansas (Cheyenne Bottoms Wildlife Area, Barton County, although summering birds and immatures have been seen in several other western counties (Finney, Seward, and Meade). There were nine confirmed nestings during breeding bird atlas surveys in South Dakota (Peterson, 1995), six in Nebraska (Mollhoff, 2001), and one in Kansas (Busby & Zimmerman, 2001).

Nest Location: Nesting is semi-colonial in water ranging from less than a foot to about three feet deep. Nests are usually on floating emergent vegetation, particularly cattail rootstalks. Muskrat houses are sometimes used, but nest substrates tend to be smaller and lower than those used by Forster's terns.

Clutch Size and Incubation Period: From two to four eggs (151 Iowa clutches averaged 2.6). The eggs are buffy to olive with extensive blackish spotting. The incubation period is 21–24 days, usually about 21 days. Single-brooded, at least in this region, but probable renesting has been reported.

Time of Breeding: North Dakota egg dates are from May 28 to July 24, and flightless young have been seen from June 21 to July 25. Nests in South Dakota have been reported from May 31 to July 19 (Peterson, 1995). Nebraska egg records are from June 4 to July 11 (Mollhoff, 2001). In Kansas, egg dates are from June 11 to August 12.

Breeding Biology: Prenesting behavior in black terns is marked by two types of display flights, including "fish flights" (the birds usually carry insects rather than fish), normally performed by two birds, and "flock flights" involving most or all of the birds of an entire nesting area. In the courtship phase one bird (probably the male) postures and calls while standing on a potential nest site. Also, the two birds make an aerial glide downward from several hundred feet while maintaining a fixed position relative to each other. Nesting sites of the previous year apparently are not reused. The nests seem to be built from materials gathered in the immediate vicinity of the nest rather than carried in. Both sexes assist in incubation, and both brood the young for at least eight days after hatching. Little brooding is done after that time, though the chicks are unable to fly until they are more than 20 days old. Young birds are fed almost exclusively with insects and continue to feed on them for a time after they fledge.

Comments: National Breeding Bird Surveys between 1966 and 2009 indicate that the species underwent a statistically significant population decline (0.6 percent annually) during that period. Partly for a sharp decline since the 1960's, no current overall recent population estimates are available (B.O.N.A. 147).

Suggested Reading: Goodwin 1960; Bailey 1977; B.O.N.A. 147; Johnsgard, 2012.

Common Tern. *Sterna hirundo*

An uncommon to rare spring and fall migrant in eastern parts of the region, and a local breeder in northeastern South Dakota.

Migration: Sixty-five initial spring sightings in Nebraska are from March 18 to June 7, with a median of May 5. Half of the records fall within the period April 24–May 15. Fourteen final spring sightings in Nebraska are from April 25 to June 6, with a median of May 11. Eleven initial fall sightings are from August 9 to October 14, with a median of September 2. Frequent confusion with the Forster's tern makes the status of this species somewhat uncertain, but migrants are most common in the northeastern portions of the Great Plains region.

Habitats: Migrants use lakes, reservoirs and rivers, and less often are found near smaller marshes and ponds. The species also breeds along coastlines on sandy beaches.

Breeding Status: Breeds locally in northeastern South Dakota. There was one confirmed nesting during breeding bird atlas surveys in South Dakota (Peterson, 1995), but none in Nebraska or Kansas.

Nest Location: Sparsely vegetated islands are preferred nesting habitats in the Great Plains; grassy uplands and sandy beach areas are used in coastal regions. Nesting is in colonies, and nests are simple scrapes, often with little or no lining. Nest sites are usually near vegetation or other upright objects and are very often in previous nest hollows or natural depressions.

Clutch Size and Incubation Period: From two to four eggs, usually three (93 North Dakota nests averaged 2.8). The eggs are buffy to cinnamon, with dark brown spotting, especially at the larger end. The incubation period is usually 24–26 days, rarely as short as 21 days. Single-brooded, with renesting typical only when the first clutch has been lost very early in the season.

Time of Breeding: Egg dates in North Dakota range from June 8 to July 28, and dependent young have been seen from June 22 to July 31. Nests in South Dakota have been reported from May 28 to July 9 (Peterson, 1995).

Breeding Biology: As terns arrive on their nesting grounds, the first birds are those that have nested there formerly, and males soon begin to establish and occupy territories. In early aerial displays, or "fish flights," small fish are exchanged among the participants, but little or no sexual recognition is likely at this stage. After sexual recognition the true courtship is under way; aerial glides replace the typical fish flights, and terrestrial displays such as parading around the potential mate or incipient nest-building or scrape-digging are typical. Copulation begins at about the time of scrape-making, and egg-laying soon follows. The egg-laying rate is rather variable, and incubation begins immediately, so that hatching is staggered. Both sexes incubate, but the females do so much more intensively and perform about three-fourths of the total incubation. The young are precocial and may leave the nest by the second day after hatching. Adults learn to recognize their young by the time they are five days old, but adults lacking chicks sometimes adopt orphan young and care for them as their own. Young birds reach flight stage at an average age of 30 days but continue to beg food for several weeks thereafter. After the young birds leave the ternery they typically do not return for three years, until they are sexually mature.

Comments: National Breeding Bird Surveys between 1966 and 2009 indicate that the species underwent a statistically significant population decline (7.4 percent annually) during that period. The North American population was estimated at about 150,000 pairs in the early 2000's (B.O.N.A. 618).

Suggested Reading: Palmer 1941; B.O.N.A. 618.

Forster's Tern. *Sterna forsteri*

A common spring and fall migrant throughout the region, and a rather localized summer resident in the Sandhills, especially Garden, Cherry and Grant counties. The species probably also breeds in some Sheridan County marshes. Breeding is regular in Minnesota and the Dakotas, and migrants appear throughout the entire Great Plains region.

Migration: Fifty-eight initial spring sightings in Nebraska are from April 11 to June 8, with a median of April 28. Half of the records fall within the period April 19 to May 5. Twenty-one final spring sightings in Nebraska are from April 24 to June 6, with a median of May 17. Thirteen initial fall Nebraska records are from July 21 to September 22, with a median of August 1. Twenty final fall sightings are from August 1 to October 8, with a median of September 11.

Habitats: Associated with lakes, rivers and marshes while on migration. Breeding occurs in large marshes having extensive areas of emergent vegetation or muskrat houses for nesting sites. It is also found around lakes, salt marshes, and coastlines but is more of a marshland species than the common tern. Small marshes seem to be avoided.

Breeding Status: Breeds uncommonly and locally in eastern and south-central South Dakota (Bennett, Clark, Codington, Day and Sully counties) and a local but regular breeder in Nebraska (Garden and Sheridan counties). In Kansas, breeding has only been reported from Cheyenne Bottoms Wildlife Area and Quivira N.W.R. (Thompson *et al.,* 2011). There were seven confirmed nestings during breeding bird atlas surveys in South Dakota (Peterson, 1995), four in Nebraska (Mollhoff, 2001), and two in Kansas (Busby & Zimmerman, 2001).

Nest Location: Nests are typically in or near water, usually on floating vegetational debris, but at times they are in a depression in sand or mud. The birds sometimes nest on small islands like common terns, but the two species rarely nest together. Studies in Iowa indicate that large muskrat houses are favored nest sites, especially when they are near the edges of open pools of water.

Clutch Size and Incubation Period: From one to four eggs (92 Iowa clutches averaged 2.5). The eggs are buffy to buffy-olive with dark brown spotting. The incubation period averages 24 days. Single-brooded, but probable renesting has been reported.

Time of Breeding: The probable period of breeding in North Dakota is late May to late July, with newly hatched young recorded in late June and early July. Nests in South Dakota have been reported from May 29 to July 13 (Peterson, 1995). Nebraska egg records are from July 11–20, and young seen as early as July 20 (Mollhoff, 2001),

Breeding Biology: Shortly after they arrive at their nesting marshes, Forster's tern pairs begin to seek out nest sites. They are relatively colonial, and as many as five nests may be placed on a favorable site, such as a large muskrat house. The floating rootstalks of cattails may also serve as a nest site, but such locations are more often used by black terns. Nest-building is initiated almost simultaneously by all members of a colony, and both sexes incubate. Wind and wave action, house building by muskrats, and possibly intraspecific hostility are probably major causes of egg loss, which seems to be relatively high in this species. Little information is available on the growth of the young, but presumably they fledge in about 3–4 weeks, as is typical of the common tern.

Comments: National Breeding Bird Surveys between 1966 and 2009 indicate that the species underwent a statistically non-significant population decline (0.8 percent annually) during that period. The North American population was estimated at about 44,000 pairs in the early 2000's, with 40,000 of these in the U.S. and Great Lakes region (B.O.N.A. 595).

Suggested Reading: Bergman, Swain, and Weller 1970; McNicholl 1971; B.O.N.A. 595; Johnsgard, 2012.

ORDER CORACIIFORMES – KINGFISHERS & RELATIVES

Family Alcedinidae–Kingfishers

Belted Kingfisher. *Ceryle alcyon*

A common spring and fall migrant and summer resident throughout the entire Great Plains region in suitable habitats, and an uncommon winter resident where open water persists.

Migration: Forty-three initial spring sightings in Nebraska range from January 2 to May 10, with a median of March 20. Half of the records fall within the period February 14 to April 10. Forty-seven final fall sightings are from July 26 to December 31, with a median of November 15. The concentration of fall Nebraska records toward the end of the year (nearly half occurring in December) suggest that the species over-winters frequently.

Habitats: Throughout the year this species occurs near wetlands supporting populations of fish, amphibians and similar aquatic life. Nests are excavated from nearly vertical earth exposures in bluffs, road cuts, eroded streambanks, and the like.

Breeding Status: A pandemic breeder in suitable habitats throughout the region. There were nine confirmed nestings during breeding bird atlas surveys in South Dakota (Peterson, 1995), 32 in Nebraska (Mollhoff, 2001) and 26 in Kansas (Busby & Zimmerman, 2001). A state-level analysis of four decade-long periods of Christmas Bird Counts (1967-68 to 2006–7) extending from North Dakota to the Texas panhandle indicated a late-December population peak in Kansas (Johnsgard and Shane, 2009).

Nest Location: The nest consists of a rather long burrow, often about five feet long but up to 15 feet, with an entrance about 3–4 inches in diameter (slightly wider than high). It is usually within three feet of the top of the bank and at least five feet above ground level. The nest cavity is an enlarged area at the end of the burrow and is often lined with disgorged food pellets. Sandy clay soil seems to be the preferred substrate, and nests are usually near a dead or dying tree.

Clutch Size and Incubation Period: From 5–8 eggs, which are white and glossy. The incubation period is 23–24 days. Single-brooded, but a persistent renester.

Time of Breeding: The nesting season in North Dakota probably extends from mid-April to late July. Nests in South Dakota have been reported from May 9 to July 20 (Peterson, 1995). Nebraska egg records are from May 24 to

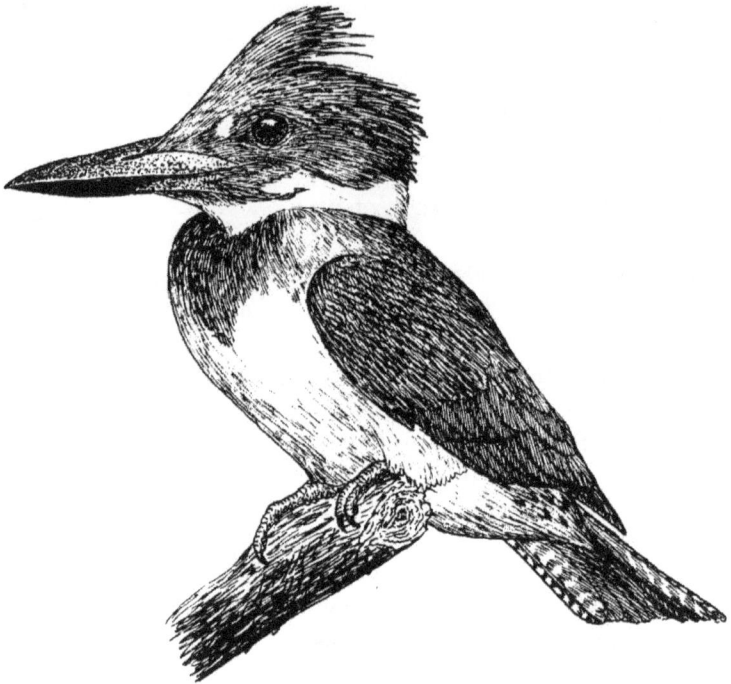

June 4 (Mollhoff, 2001). In Kansas, eggs have been reported in early May, and Oklahoma egg dates are from April 26 to July 1, with nestlings reported from June 4 to July 15.

Breeding Biology: Belted kingfishers take up residence in suitable habitats that allow for large home ranges. At times they may forage up to five miles from the nest site, and a population density of about one pair per 1.8 square miles of habitat has been estimated at Lake Itasca in Minnesota. Small fish averaging about 3–4 inches long compose more than half their diet. Both sexes participate in nest excavation, which may require up to three weeks. Both sexes incubate, apparently beginning after the last egg is laid. After hatching the male also assists in getting food. At night the male usually roosts away from the nest, sometimes in a separate burrow or in a forested area. The young are relatively helpless and spend much time clinging to one another, apparently to maintain body warmth. They remain in the nest for at least a month, when they are first able to fly, then stay near it for the next few days while their parents teach them to catch fish. The adult captures a fish, beats it until it is nearly senseless, and drops it back into the water. The young are encouraged to capture these easy prey, and gradually learn to catch normal fish. Within ten days they are relatively independent and soon leave the vicinity of the adult pair.

Comments: The belted kingfisher is Nebraska's only representative of this large and diverse family of mostly fish-eating birds, although the smallest kingfishers are largely insectivorous and the largest ones are omnivorous. Only the female of this species has chestnut underparts, making the sexes easy to identify. National Breeding Bird Surveys between 1966 and 2009 indicate that the species underwent a statistically significant population decline (1.5 percent annually) during that period.

Suggested Reading: White 1953; Cornwell 1963; B.O.N.A. 84.

ORDER PASSERIFORMES – PERCHING BIRDS

Family Hirundinidae–Swallows

Northern Rough-winged Swallow. *Stelgidopteryx serripennis*

A common spring and fall migrant and summer resident throughout virtually the entire Great Plains.

Migration: The range of 136 initial spring sightings in Nebraska is from March 2 to May 29, with a median of April 28. Half of the records fall within the period April 18 to May 6. Seventy-two final fall sightings are from July 21 to October 15, with a median of September 3. Half of the records fall within the period August 23–September 15.

Habitats: This is an open–country species, often found near rivers or creeks having exposed vertical banks of clay or other materials that can be excavated to provide nest sites. Unlike the colonial bank swallow, this species is a solitary nester, but in common with the bank swallow, its local distribution is dependent on nesting sites.

Breeding Status: Breeds virtually throughout the region, but variably abundant. Generally common in the eastern parts of the Dakotas but uncommon in western Nebraska. There were 23 confirmed nestings during breeding bird atlas surveys in South Dakota (Peterson, 1995), 87 in Nebraska (Mollhoff, 2001) and 135 in Kansas (Busby & Zimmerman, 2001).

Nest Location: Nests are excavated in banks of clay, sand, or gravel. They are much like those of the bank swallow but are solitary rather than in colonies. Natural rock crevices, fissures, and even drainpipes are sometimes also used. Unlike bank swallows, the birds do not use feathers for a nest lining, and their nest is much bulkier.

Clutch Size and Incubation Period: From 48 white eggs, usually 5–6. The incubation period is 15–16 days. Single-brooded, but renesting efforts are frequent.

Time of Breeding: Dates of active nests in North Dakota range from May 10 to July 15, with eggs seen as late as June 26. Nests in South Dakota have been reported from May 13 to July 24 (Peterson, 1995). Nebraska egg records are from June 15–20 (Mollhoff, 2001). In Kansas, egg dates are from May 11 to June 30 with most eggs laid between May 21 and June 10. Nest-building has been seen from May 18 to June 10, and nestlings have been seen from June 11 to July 20.

Breeding Biology: Almost as soon as they arrive on their nesting grounds, these swallows begin to show interest in suitable nesting sites, and they may seek out old kingfisher or bank swallow excavations that are still usable. Males establish a limited territory around a potential nest site, perching near it and pursuing females from it. Females carrying nesting materials are especially pursued, although this behavior may be associated more with copulation than with courtship. Copulation has not been described and presumably occurs in the nesting cavity. Evidently only the female gathers and carries nest-lining material; apparently neither bird does any excavating. An average of about six days is needed to construct the rather bulky nest, but it may take as long as 20 days. The female usually starts incubating with the laying of the next-to-the-last egg, and hatching may extend a few hours or as long as several days. Brooding is done primarily if not exclusively by the female, but both sexes feed the young. The young birds are able to fly some days before they leave the nest, which usually occurs at 18–21 days of age. Young birds rarely return to the nest after they leave it, and there is no evidence on how long the young remain dependent on their parents for food after fledging.

Comments: Rough-winged swallows are among the most common swallows in eastern Nebraska, especially near streams having steep-sided mud banks. They also nest in natural cavities of rocky outcrops, although perhaps less frequently. They have even been found to use horizontally installed drainpipes as nesting cavities. National Breeding Bird Surveys between 1966 and 2009 indicate that the species underwent a statistically non-significant population decline (0.4 percent annually) during that period. Its North American population (north of Mexico) has been estimated at about five million birds (Rich *et al.,* 2004).

Suggested Reading: Lunk 1962; Bent 1942; B.O.N.A. 234.

Bank Swallow. *Riparia riparia*

A common spring and fall migrant in eastern Nebraska and South Dakota, breeding south to northeastern Kansas, and is a regular migrant farther south.

Migration: The range of 104 initial spring sightings in Nebraska is from March 20 to June 8, with a median of May 6. Half of the records fall within the period April 28–May 6. Sixty-five final fall sightings range from July 31 to October 29, with a median of September 8. Half of the records fall within the period August 23–September 15.

Habitats: While in Nebraska this species occurs in a variety of open habitats, especially grasslands and croplands, but is typically found near water and is dependent on suitable potential nest sites in the form of vertical banks of clay, sand or gravel that can be excavated by the birds.

Breeding Status: Breeds in suitable habitats over the northern half of the region, being rather common from South Dakota southward through Nebraska. There were 27 confirmed nestings during breeding bird atlas surveys in South Dakota (Peterson, 1995), 39 in Nebraska (Mollhoff, 2001) and 14 in Kansas (Busby & Zimmerman, 2001).

Nest Location: Nests are invariably in vertical banks of clay, sand, or gravel; the bird is characterized by colonial rather than solitary nesting. The openings are near the top of the bank and are about 11/2 by 2 ¼ inches. The burrows average two feet in length and are turned slightly upward. The nest is a platform of vegetation, usually with feathers for lining.

Clutch Size and Incubation Period: From 4–6 white eggs, usually five (60 Kansas clutches averaged 4.8; six North Dakota clutches averaged 5.3). The incubation period is 15 days. Normally probably single-brooded, but two broods were reported in one study.

Time of Nesting: North Dakota egg dates range from June 5 to July 5. Nests in South Dakota have been reported from May 20 to August 9 (Peterson, 1995). Nebraska egg records are from May 18 to June 2 (Mollhoff, 2001). Kansas records extend from May 11 to June 20, with most of the clutches laid between May 21 and June 10.

Breeding Biology: Shortly after bank swallows arrive in a nesting area, they begin to gather near the breeding site. Unpaired birds (a male in at least one determined instance) apparently select a burrow site, which may be the same burrow they used the previous year. Thereafter they defend the area from intrusion, although potential mates continue to return to a defended spot until one is eventually tolerated and accepted. Sexual chases of the female by the male are a common feature of pair-formation, accompanied by male song. Another vocalization, the mating song, is uttered by both members of a pair as they sit side by side or facing each other in the burrow opening. This behavior may be a preliminary to copulation, which probably occurs in the nest chamber. When a burrow needs to be dug or deepened, both sexes share equally in the task, then gather materials such as feathers and grass for nest lining. The incubation is by the female, and may begin before the clutch is completed. Thus some eggs may hatch as early as 13 days after the clutch has been completed. Both parents alternate at brooding the young, and both feed the young and keep the nest clean. Birds as young as 20 days of age may be able to fly but often do not leave the nest for some time thereafter; or they may return to their burrows after initial flights.

Comments: Bank swallows are local nesters in the region, requiring rather large areas of barren road-cuts to support a breeding colony. The nearly vertical clay-like banks of loess that are common beside roads along the Missouri Valley provide a perfect nesting situation for these birds. National Breeding Bird Surveys between 1966 and 2003 indicate that the species underwent a statistically significant population decline (5.2 percent annually) during that period. Its North American population (north of Mexico) has been estimated at about 14 million birds (Rich *et al.,* 2004).

Suggested Reading: Bent 1942; Peterson 1955; B.O.N.A. 414.

Cliff Swallow. *Petrochelidon pyrrhonota*

A common spring and fall migrant and summer resident through the entire Great Plains region.

Migration: The range of 125 initial spring sightings in Nebraska is from March 22 to June 10, with a median of April 28. Half of the records fall within the period April 29 to May 18. The range of 101 final fall sightings is from July 22 to October 30, with a median of September 4. Half of the records fall within the period August 20 to September 15.

Habitats: This species occurs over open areas of farmlands, towns, near cliffs, around bridges, and in other areas where mud supplies and potential nest sites exist on vertical and overhanging surfaces. This species is highly colonial, typically nesting in the same locations year after year, and re-using old nests if they are present.

Breeding Status: Pandemic, breeding throughout the region in suitable habitats. There were 113 confirmed nestings during breeding bird atlas surveys in South Dakota (Peterson, 1995), 136 in Nebraska (Mollhoff, 2001) and 155 in Kansas (Busby & Zimmerman, 2001).

Nest Location: Nests are gourd–like structures of dried mud, attached to the vertical and overhanging surfaces of cliffs, buildings, bridges, and other structures. The nest has a tubular rounded entrance on the lower side and is lined on the inside with vegetation, but rarely feathers. Nesting is strongly colonial, with up to several hundred nests often occurring in favorable locations such as concrete bridges, rarely a thousand or more.

Clutch Size and Incubation Period: From 36 eggs, usually 4–5. The eggs are white with brown spotting. The incubation period is 15 days.

Time of Nesting: North Dakota egg dates are from June 13 to July 9, although active nests have been reported from May 8 to September 6. Nests in South Dakota have been reported from May 12 to August 7 (Peterson, 1995). Nebraska egg records are from May 25–28, and young have been seen as early as July 4 (Mollhoff, 2001). Kansas egg dates are from May 21 to June 30, with most clutches laid between May 21 and June 10. Young in the nest have been reported May 29 to August 17, and newly fledged young as early as June 6.

Breeding Biology: At least in the northern states, cliff swallows begin to pair immediately upon arrival of their nesting grounds. This activity takes place at or near the nest, and the pair bond apparently consists primarily of mutual tolerance at the nesting site. Male "primary squatters" persistently return to specific perching places, and their singing attracts secondary visitors to that location, some of which are unpaired females. Both sexes defend the nest site, and both bring mud to construct the nest, which requires nearly two weeks of effort. When the nest is nearly completed, copulation occurs in the nest cup, and copulatory behavior continues until the middle of the laying period. Many cliff swallows occupy old nests if they are still usable; otherwise they construct entirely new ones. The incubation may begin before the clutch is complete, and males regularly participate. There is a relatively long nestling period in this species, averaging about 24 days, and a relatively low proportion of females (27 percent in one study) attempt a second clutch. In at least some cases females change mates for their second nesting, and a considerable amount of courtship activity is evident between broods. There is also a high incidence of stolen copulations involving unmated pairs, and of females depositing their eggs into the nests of others.

Comments: There are few if any locations in America that support larger numbers of cliff swallows than the bridges, culverts, and similar structures in the central and western Platte Valley. Breeding colonies supporting up to several thousand nests are present in a few locations, and the numbers of mosquitoes, midges and similar aerial insects that are consumed in a summer must be astronomical. National Breeding Bird Surveys between 1966 and 2009 indicate that the species has had a statistically non-significant population increase (0.7 percent annually) during that period. Its North American population (north of Mexico) has been estimated at about 80 million birds (Rich *et al.*, 2004).

Suggested Reading: Emlen 1954; Samuel 1971; B.O.N.A. 149.

Family Troglodytidae–Wrens

Sedge Wren. *Cistothorus platensis*

An uncommon spring and fall migrant and local summer resident in eastern parts of South Dakota, Nebraska and Kansas.

Migration: Twenty-five initial spring sightings in Nebraska range from April 16 to June 3, with a median of May 8. Half of the records fall within the period May 1–12. Seventeen final fall sightings are from July 29 to October 22, with a median of September 28. Half of the records fall within the period September 11 to October 9. Many birds seem to arrive in mid-summer and begin nesting at that time.

Habitats: In Nebraska, and the northern plains these birds breed in wet meadows, typically those dominated by sedges and tall grasses, and less often breed in the emergent vegetation of marshes as well as retired croplands and hayfields.

Breeding Status: Breeds southward through eastern South Dakota, to central Nebraska. It is an occasional summer resident in the eastern third of Nebraska, mostly east of a line from Knox to Gage counties, but west to at least Hall County and perhaps to Lincoln County (Mollhoff, 2001). In Kansas, it is rare and irregular, with breeding records limited to a few eastern counties. There were no confirmed nestings (but 22 probable nestings) during breeding bird atlas surveys in South Dakota (Peterson, 1995), two in Nebraska (Mollhoff, 2001), and one confirmed nesting in Kansas (Busby & Zimmerman, 2001).

Nest Location: Nests are constructed over land or water in dense growing vegetation and are usually 1–3 feet above the substrate. They are globular structures about four inches in diameter, with a lateral entrance above the equator, lined with plant down, hair, or similar materials.

Clutch Size and Incubation Period: From 4–8 white eggs, often seven. The incubation period is 12–14 days. Reportedly double-brooded, but single-brooding was observed in an Iowa study.

Time of Breeding: North Dakota egg dates are from June 7 to August 10. Nests in South Dakota have been reported from June 20 to July 9 (Peterson, 1995). Nebraska egg records are from August 24 to September 1, with young seen as early as August 28 (Mollhoff, 2001). Kansas egg dates are from August 10 to September 24, and newly fledged young seen as early as August 30. Apparently this species is a notably late nester, and birds arriving in mid-July are not necessarily ones that had unsuccessfully attempted nesting earlier.

Breeding Biology: Although this species is not nearly so well studied as the long-billed marsh wren, it is known that males regularly build numerous "dummy" nests, and thus a comparable pattern of pair-formation and evidently polygyny prevails. In favored habitats such as large meadows, the birds concentrate min high densities; about 35–40 singing males were counted in a Michigan meadow of only ten acres. At the peak of the nesting period the male may spend as much as 22 hours a day singing, generally from 6–12 songs per minute. When a pair bond has formed the female selects or initiates a brood nest, which tends to be lower and harder to find than the courting nests. The female does all the incubating and most of the feeding of the young, with only occasional visits by the male. The young remain in the nest about 13 days, and presumably a second brood is often initiated shortly after the first one fledges.

Comments: Sedge wrens are common breeders in southeastern Nebraska, where they arrive and begin to sing in early May. Then, in July or August a new song cycle begins, leading some to speculate that these are late arriving birds, perhaps from farther north, where they may have been unsuccessful breeders. Apparently this species is a notably late nester, and birds arriving in mid-July are not necessarily ones that attempted nesting earlier but unsuccessfully (Thompson *et al.*, 2011). National Breeding Bird Surveys between 1966 and 2009 indicate that the species had a statistically non-significant population increase (1.4 percent annually) during that period. Its North American population has been estimated at about 6.5 million birds (Rich *et al.*, 2004).

Suggested Reading: Walkinshaw 1935; Crawford 1977; B.O.N.A. 572; Johnson & Igle, varied dates.

Marsh Wren, *Cistothorus palustris*

A common spring and fall migrant throughout the region and locally common summer resident in South Dakota and Nebraska north of the Platte River, with local or infrequent breeding south of the Platte River *(Nebraska Bird Review* 39:47). There were ten confirmed nestings during breeding bird atlas surveys in South Dakota (Peterson, 1995), 25 in Nebraska, and none in Kansas (Busby & Zimmerman, 2001).

Migration: Seventy-eight initial spring sightings in Nebraska range from March 13 to June 9, with a median of May 5. Half of the records fall within the period April 26 to May 15. Thirty-two final fall sightings are from August 9 to November 22, with a median of October 2. Half of the records fall within the period September 8 to October 10. A state-level analysis of four decade-long periods of Christmas Bird Counts (1967-68 to 2006–7) extending from North Dakota to the Texas panhandle indicated a late-December population peak in northwestern Texas (Johnsgard and Shane, 2009).

Habitats: During the breeding season these birds are primarily found in freshwater marshes having extensive tall emergent vegetation, such as bulrushes and cattails. They also nest along the banks of slowly flowing brackish tidal marshes.

Breeding Status: Breeds throughout the eastern half of South Dakota, and most of Nebraska north of the Platte River, with local or infrequent breeding south of the Platte *(Nebraska Bird Review* 39:74). It also breeds locally in northeastern Kansas (Doniphan County, possibly others). There were ten confirmed nestings during breeding bird atlas surveys in South Dakota (Peterson, 1995), 26 in Nebraska (Mollhoff, 2001) and none in Kansas (Busby & Zimmerman, 2001). Until the modification of Kingsley Dam for hydropower generation in 1982, breeding occurred regularly on Lake Ogallala/Keystone.

Nest Location: Most nests are 3–5 feet above the marsh substrate, with early nests being lower and later ones higher. The nests are built in emergent vegetation (cattails, bulrushes, etc.) and are domed elliptical structures about seven inches high and five inches across, with a lateral opening about 1 ½ inches in diameter just above the equator. They are constructed of grass strips and lined with cattail down. In North Dakota, water depth at 19 nests averaged 12 inches, and 23 nest entrances averaged 16 inches above the water.

Clutch Size and Incubation Period: From 3–7 eggs (23 North Dakota clutches averaged 4.7). The eggs are cinnamon to brown with darker spots. The incubation period is 13 days. Double-brooded.

Time of Breeding: North Dakota egg dates are from May 26 to August 10, and nestlings have been reported from June 16 to August 6. Nests in South Dakota have been reported from June 1 to July 17 (Peterson, 1995). Nebraska egg records are from June 6 to July 11 (Mollhoff, 2001). Kansas egg dates are from May 30 to July 4, and newly fledged young have been seen from August 21–31.

Breeding Biology: Shortly after they arrive at their breeding areas males establish territories, which they advertise by persistent singing from all parts and by aerial displays above them. After a territory has been established, each male begins to build a number of "courting nests" (up to five or more), which are complete except for a lining. When a female selects a male as a mate she either accepts one of these nests for breeding and lines it or begins a new one, which is chiefly constructed by the male. After egg-laying has begun, the male moves to a new area in his territory and begins to advertise for additional mates. He may obtain as many as three mates, each of which incubates alone but is fed by the male. The male's role in feeding the young is small or nil; instead, he continues to maintain the territory. The young leave the nest when about 14–16 days old, but may be fed by adults for nearly two more weeks. Nest-building by the male increases during the nestling and fledgling period, in preparation for a second clutch.

Comments: The marsh wrens of Nebraska pose a problem in evolution, with two distinct song types occurring in the state, as noted above. Perhaps the two types represent "sibling species" that seem to differ only in their vocalizations, but act biologically as distinct species. Both eastern and western song types occur in Nebraska, the dividing line passes southeast through O'Neill, approximately along the Elkhorn Valley (*Nebraska Bird Review* 64:99), or along the eastern edge of the Sandhills (Mollhoff, 2001). National Breeding Bird Surveys between 1966 and 2004 indicate that the species underwent a statistically significant population increase (2.0 percent annually) during that period. Its North American population has been estimated at about 7.7 million birds (Rich *et al.,* 2004).

Suggested Reading: Verner 1965; B.O.N.A. 308.

Family Cinclidae - Dippers

American Dipper, *Cinclus mexicanus*

A permanent resident in the Black Hills, but not reported breeding from elsewhere in the region, but an extremely rare migrant in Nebraska.
Migration: Probably an irregular migrant in Nebraska. Records exist for May, June, October and December. Individuals have been reported once or more from at least Adams, Chase, Cherry, Dawes, Holt, and Sioux counties.

Breeding Status: Limited to the Black Hills of South Dakota, where it is an uncommon permanent resident.

Habitats: Throughout the year this species is normally associated with rapidly flowing mountain streams in wooded areas, but in winter the birds sometimes move to open water at lower elevations. In the Black Hills it is most common in Spearfish Canyon, but it also occurs elsewhere in the Hills.

Nest Location: Nests are usually over water, either under bridges or under overhanging rock ledges. They have also been found among the roots of fallen trees, but in all cases they are made of woven mosses, usually with a roof, sides, and a front entrance. The nest may be from 8–12 inches in external diameter, thus being rather conspicuous, or may appear to consist of a small hole about two by three inches in a vertical wall of moss on the side of a cliff. The nest cup is of coarse grass, which effectively resists moisture.

Clutch Size and Incubation Period: From 3–6 white eggs, usually four or five. The incubation period is 15–17 days, averaging 16. Single-brooded.

Time of Nesting: Nests in South Dakota have been reported from April 13 to July 5 (Peterson, 1995). Fledged young have been seen as early as June 13.

Breeding Biology: Dippers are very sedentary birds, and pairs tend to remain well separated. Studies in Montana indicate that by November the birds begin to establish winter territories, which are strongly defended through February and may include from 50 yards to as much as a half mile of stream. In spring the birds abandon these territories and begin to move upstream into breeding territories. During winter the birds also begin to sing, and singing increases in intensity to a peak in April. The songs of the two sexes are indistinguishable, and loud singing as well as wing-quivering and chasing behavior accompany pair-formation. Both sexes participate in nest-building or in reconstructing an old nest, which seems to be more common than building entirely new nests. The incubation is entirely by the female, but the male frequently brings food to her during incubation. The young are hatched with a coating of down but grow relatively slowly, so that the nestling period is surprisingly long, about 19–25 days. The female broods regularly for about a week after hatching, and males rarely or never enter the nest. Instead, the older young poke their heads out the nest entrance to be fed. When the young birds leave the nest they are nearly as large as their parents and easily flutter out to a safe landing below. After the birds leave the nest, one or both of the parents typically removes the nest lining, presumably to prepare the nest for use in another year. The young birds soon learn to clamber about on the wet rock surfaces, and they remain in the vicinity of the nest for up to 15 days after fledging. It is likely that adult birds return to their same nesting areas each year, but in at least one case an adult bird had different mates in two successive years.

Comments: American dippers are sometimes called "water ouzels," and they are the only American songbirds that regularly dive into the water and search for food at the bottom of fast-moving steams. It seems possible that some of the clear and rapidly flowing branch streams of the Niobrara River might support breeding pairs of these remarkable birds. National Breeding Bird Surveys between 1966 and 2009 indicate that the species had a statistically non-significant population decline (0.2 percent annually) during that period. Its North American population has been estimated at about 600,000 million birds (Rich *et al.,* 2004).

Suggested Reading: Hahn 1950; Bakus 1959; B.O.N.A. 229.

Family Parulidae - Wood Warblers

Prothonotary Warbler, *Protonotaria citrea*

An uncommon or occasional spring and fall migrant in northeastern Kansas and southeastern Nebraska, and a local summer resident at least as far north as Sarpy County, but with possible if unlikely breeding reported west in the Niobrara River valley to Cherry County.

Migration: Thirty-six initial spring sightings in Nebraska range from April 19 to May 24, with a median of May 12. Half of the records fall within the period May 5–17. Four final fall sightings range from July 26 to October 4, with a mean of September 11.

Habitats: This species is restricted during summer to moist bottomland forests and wooded swamps or periodically flooded woodlands, in the vicinity of running water or pools. During migration birds may also appear in other wooded areas.

Breeding Status: A rare breeder in the Missouri Valley of southeastern Nebraska (extending north to Sarpy County) southward through eastern Kansas (breeding records exist for Doniphan, Douglas, Linn, and Cowley counties). Mollhoff (32001) reported a confirmed breeding for Sarpy County, and probable breedings for Nemaha, Gage and (remarkably) Cherry counties during Nebraska breeding bird surveys. There were no confirmed nestings during breeding bird atlas surveys in South Dakota (Peterson, 1995), two in Nebraska (Mollhoff, 2001) and 14 in Kansas (Busby & Zimmerman, 2001).

Nest Location: Nests are in natural cavities or in old woodpecker holes, 5–20 feet high, often over water. Both tree stumps and trees are used, and there have been some records of nesting in birdhouses, gourds, and even tin cups. The cavity is lined with a variety of grasses, leaves, rootlets, and twigs, forming a deep hollow.

Clutch Size and Incubation Period: From 3–6 eggs (15 Kansas clutches averaged 4.5). The eggs are white with extensive brown spotting. The incubation period is 12–14 days. Frequently double-brooded, at least in southern regions.

Time of Breeding: Nebraska egg records are from May 29 to June 17 (Mollhoff, 2001). Kansas egg records are from May 19 to July 1, with a peak in mid-June. Young in the nest have been seen there from June 9 to July 11, and recently fledged young from June 9 to July 23. In Oklahoma, breeding records are from May 5 (nest-building) to early August (dependent young).

Breeding Biology: As soon as the males return to their breeding areas in spring, they become territorial and particularly try to stake out areas along riverbanks or other well-shaded water edges. The presence of suitable nest sites, preferably downy woodpecker holes only 5–6 feet above the water or ground, is critical, and males begin to carry moss into the cavities even before the females arrive. Apparently it is not typical for these birds to mate with the partner from the previous year. The female completes the nest and performs all the incubation. Fledging occurs after 10–11 days, and a second clutch may be begun in the same nest.

Comments: This is another species that occurs locally as far north as Nebraska's mature riparian forests of Fontenelle Forest and evidently nests there. There is also an early breeding record for Otoe County. It is much more common in Kansas, where there are breeding records for 20 counties. National Breeding Bird Surveys between 1966 and 2009 indicate that the species underwent a statistically significant population decline (1.0 percent annually) during that period. Its North American population has been estimated at about 1.8 million birds (Rich *et al.*, 2004).

Suggested Reading: Walkinshaw 1953; Bent 1953; Curson *et al.*, 1995; B.O.N.A. 408.

Louisiana Waterthrush, *Parksia motacilla*

A rare to uncommon spring and fall migrant in northeastern Kansas and eastern Nebraska, becoming rarer to the west and north, and reported from as far west in Nebraska as Lincoln and Garden counties. Breeding is regular in eastern Kansas.

Migration: Seventy-six initial spring Nebraska sightings range from March 30 to May 29, with a median of May 8. Half of the records fall within the period May 2–14. Ten final spring sightings are from April 29 to May 29, with a median of May 15. Ten fall sightings are from July 29 to September 24, with a median of August 29.

Habitats: The species is usually found near swift streams, in wooded and hilly country, but also breeds around lagoons and swamps where trees grow to the shoreline.

Breeding Status: Breeds from the Kansas River Valley of Kansas (west to Riley, Butler and Cowley counties), and south along the eastern edge of the state to Oklahoma. There are scattered western breeding records for Comanche, Decatur, and Kiowa counties. Confirmed breeding in Nebraska during the atlasing years was limited to Washington and Richardson counties (Mollhoff, 2001). Breeding has recently been suggested for Fontenelle Forest and Neale Woods, both near Omaha, and Platte River State Park, Cass County (Loren Padelford, pers. comm.), and singing birds have been reported north to Dakota County.

Nest Location: Nests are usually close to water, either in holes in steep banks or under the roots of nearby trees, well hidden by overhanging vegetation. The nest is made of dead leaves that are well packed and intermixed with twigs, and the nest cup is lined with grasses, rootlets, and so forth. There may be a pathway of leaves leading to the nest.

Clutch Size and Incubation Period: From 4–6 eggs, white with extensive spotting or blotching of gray and brown. The incubation period is 12-14 days. Single-brooded.

Time of Breeding: A Nebraska egg record is from May 18, and fledged young have been seen as early as June 12 (Mollhoff, 2001). In Kansas, eggs are found from May 9 to June 8, young in the nest from May 18 to July 17, and newly fledged young as early as 31 May. In Oklahoma fledglings have been seen as early as May 12, while nestlings have been noted as late as June 25. An early breeding and fall departure appeared typical of this species.

Breeding Biology: Males establish territories immediately after arriving in the spring and begin to advertise them from song posts in trees. Territories are long and narrow, centered on fast-flowing streams, and pairs typically occupy about 200-400 yards of stream habitat. The nest is constructed by both sexes, with the female doing most of the work. The main structure takes a day or two, and 2–3 more days are needed for lining it and for laying the first egg. The female incubates alone, and the male rarely is seen near the nest until the young are about to hatch, when he may begin to bring food. The young leave the nest when ten days old, and within another month they begin to wander about unattended by the adults.

Comments: This waterthrush may be expanding its breeding range in eastern Nebraska; at least it is being reported during summer more frequently in recent years. National Breeding Bird Surveys between 1966 and 2009 indicate that the species had a statistically non-significant population increase (0.4 percent annually) during that period. Its North American population has been estimated at about 26.000 birds (Rich *et al.,* 2004).

Suggested Reading: Bent 1953; Eaton 1958; B.O.N.A. 151

Common Yellowthroat, *Geothlypis trichas*

A common to abundant spring and fall migrant and summer resident \ throughout the Great Plains region.

Migration: The range of 107 initial spring sightings in Nebraska is from April 5 to June 10, with a median of May 7. Half of the records fall within the period May 2–13. The range of 114 final fall sightings is from July 20 to October 29, with a median of September 13. Half of the records fall within the period August 30 to October 3.

Habitats: This species is primarily found near moist ground, at aquatic sites, and among associated lush vegetation, including tall grasses and often shrubs and small trees. Pond or river margins, or swampy areas, are especially utilized, but occasionally it is found in upland thickets of shrubs and small trees, in poorly tended orchards, retired croplands, or weedy residential areas.

Breeding Status: Pandemic, breeding throughout the region in suitable habitats. There were 20 confirmed nestings during breeding bird atlas surveys in South Dakota (Peterson, 1995), 31 in Nebraska (Mollhoff, 2001) and 38 in Kansas (Busby & Zimmerman, 2001).

Nest Location: Nests are near or, rarely, on the ground, in heavy growth of weeds, grasses, or low shrubs generally in vegetation less than two feet high. They are bulky, made of coarse grasses, leaves, and sometimes mosses and lined with grasses, bark fibers, and hair. Nests are well concealed unless the vegetation is parted, and are usually no more than a few inches above the ground, rarely as high as three feet.

Clutch Size and Incubation Period: From 3–6 eggs, normally four. The average of 16 first nestings in Minnesota and Michigan was 3.9 eggs, and eight subsequent nestings averaged 3.8. The eggs are white, with brown speckling near the larger end. The incubation period is 12 days. Frequently double-brooded in favorable habitats, and known to be a persistent renester.

Time of Breeding: In North Dakota egg dates are from mid-June to mid-July, and nestlings have been seen as early as June 6. Nests in South Dakota have been reported from June 9 to July 11 (Peterson, 1995). Nebraska egg records are from June 8–21 (Mollhoff, 2001). In Kansas, egg records are from May 29 to June 25, young in the nest from June 6–16, and recently fledged young from June 29 to July 10.

Breeding Biology: As soon as they arrive in spring, males establish territories that usually are less than two acres in area, but in some instances may exceed three acres (one bigamous male was found to occupy an area of 3.4 acres). Nests are usually built in the drier and more open parts of the territory, but water is always nearby. The female builds the nest over a period of two to five days and also performs all the incubation. Feeding of the young is done by both sexes, and the young may leave the nest when only 7–8 days old. However, they cannot fly until they are 11–12 days, old and they do not begin feeding on their own until about three weeks old. One or both parents tend them until they are 4–5 weeks old. Most or all females attempt to raise second broods, but few are successful. At least some females built at least three nests, and mate-changing between broods apparently is infrequent.

Comments: This species' persistently repeated *"whichity"* notes emanate from weedy or shrubby thickets and reveal the species' identity long before the bird itself is generally seen. National Breeding Bird Surveys between 1966 and 2009 indicate that the species had a statistically significant population decline (1.0 percent annually) in that period. Its North American population has been estimated at about 32 million birds (Rich *et al.,* 2004).

Suggested Reading: Stewart 1953; Hofslund 1959; Curson *et al.,* 1995; B.O.N.A. 448.

Family Emberizidae – Sparrows

Le Conte's Sparrow, *Ammospiza leconteii*

A rare to uncommon summer resident of northeastern South Dakota, and an uncommon spring and fall migrant in eastern parts of Nebraska and Kansas, rare or absent westwardly.

Migration: Fifty-four initial spring Nebraska sightings are from April 1 to June 7, with a median of April 29. Half of the records fall within the period April 21 to May 8. Thirteen final spring sightings are from April 17 to May 19, with a median of May 2. Twenty-one initial fall sightings are from July 25 to October 15, with a median of September 22. Seventeen final fall sightings are from July 26 to November 9, with a median of October 20. Sharpe, Silcock and Jorgensen (2001) noted that spring reports extend from March 30 to May 19, and fall reports from September 14 to October 30. There are also a few summer records for Nebraska .

Habitats: The prime breeding habitat of this species consists of hummocky alkaline wetlands (fens), but the species also occurs less commonly in tall-grass prairie, the wet meadow zone of prairie ponds or lakes, and domestic hayfields or retired croplands. Migrants are found in wet meadows and marshy edges with sedges, cattails, and tall grasses. There were no confirmed nestings (but one probable and two possible nestings) during breeding bird atlas surveys in South Dakota (Peterson, 1995), none in Nebraska (Mollhoff, 2001), and none in Kansas (Busby & Zimmerman, 2001).

Nest Location: Nests are placed on the ground in dense herbaceous vegetation, usually in the drier border areas of wetlands where vegetation is luxuriant. Nests in North Dakota are often built in cordgrass. The nest is built of grasses woven among standing plant stems and thus often is elevated slightly above the ground, and it is lined with fine grasses.

Clutch Size and Incubation Period: From 3–5 eggs, usually four. The eggs are white, with spots, dots, and blotches of brown, rather evenly distributed. The incubation period is 11–13 days. Probably single-brooded, but known to renest.

Time of Breeding: In North Dakota the probable breeding season is from late May to mid-August, with a peak in early June to late July. Egg dates are from May 30 to July 21.

Breeding Biology: This elusive marshland-adapted species has been described as "mouselike" in behavior, and its territorial advertisement song sounds more like that of a grasshopper than a bird. Like the grasshopper sparrow, it also has a more prolonged and repeated song that is less frequently heard, and it sometimes sings in flight. Territories are established, but they rarely overlap, and territorial interactions are rarely seen. Incubation and brooding are by the female alone; presumably the male helps feed the young.

Comments: National Breeding Bird Surveys between 1966 and 2009 indicate that the species underwent a statistically non-significant population decrease (1.2% annually) during that period. Its North American population has been estimated at about 2.9 million birds (Rich *et al.,* 2004).

Suggested Reading: Bent 1968; Murray 1969; B.O.N.A. 224; Johnson & Igle, varied dates.

Nelson's Sparrow, *Ammospiza caudacuta*

A rare to uncommon summer resident of northeastern South Dakota, and an inconspicuous and seemingly rare spring and fall migrant in eastern parts of Nebraska and Kansas. Most of the Nebraska sightings are from the eastern half of the state, but it has been seen west to Custer, Cherry, and Sheridan counties. It has been reported more often in Lancaster County than elsewhere in the state, probably as a reflection of birding intensity rather than habitat availability.

Migration: Five spring Nebraska sightings are from March 29 to May 30, with a mean of May 5. Three of the records are for the month of May. Nine fall records are from September 7 to October 21, with a mean of September 30. Five of the records are for the month of October. Sharpe, Silcock and Jorgensen (2001) noted that spring reports extend from April 26 to May 30, and fall reports from September 7 to November 7.

Habitats: In our region this species is associated primarily with alkaline, hummocky bogs (fens) and the marshy zones of prairie lakes and ponds during years when water levels are low. Less often, wet meadow zones are used. Migrants are found along the wet edges of marshes and sloughs, usually in even wetter habitats than those used by the Le Conte's sparrow. There was one confirmed nesting during breeding bird atlas surveys in South Dakota (Peterson, 1995), none in Nebraska (Mollhoff, 2001), and none in Kansas (Busby & Zimmerman, 2001).

Nest Location: Nests are on the ground, usually sunken to the ground level but sometimes built among upright stems and thus elevated in thick clumps of grass. The nests are constructed of dry grasses and lined with finer grasses.

Clutch Size and Incubation Period: From 36 eggs. The eggs are pale greenish white with brown dots and spots. The incubation period is 11 days. Frequently double-brooded.

Time of Breeding: The probable breeding season in North Dakota is from early June to late August, with a peak from mid-June to early August. Egg dates range from June 12 to July 12.

Breeding Biology: In contrast to the related Le Conte sparrow, males of this species are non-territorial and simply occupy home breeding ranges. Furthermore, the birds tend to be somewhat colonial and are evidently promiscuous. The nonterritorial and semicolonial nature of this species may enable females to locate males, since the males' songs are weak and uttered relatively infrequently. The songs, which include a flight song, probably serve as an index of sexual excitement, since they do not advertise territories. Males take no role in nesting activities. The young remain in the nest for 9-10 days and continue to be fed for about 20 days after leaving the nest.

Comments: Even harder to see than the Le Conte's sparrow, the Nelson's sparrow is also a bird of dense, marshy vegetation. Breeding Bird surveys between 1966 and 2009 indicate that the species underwent a nonsignificant population increase (1.2% annually) during that period. Its North American population has been estimated at about 250,000 birds (Rich *et al.,* 2004).

Suggested Reading: Woolfenden 1956; Murray 1969; B.O.N.A. 112; Johnson & Igle, varied dates.

Swamp Sparrow, *Melospiza georgiana*

A spring and fall migrant throughout the region. Breeding occurs very locally in eastern South Dakota and Nebraska, especially the Sandhills.

Migration: Thirty-three initial spring sightings in Nebraska are from March 30 to June 6, with a median of April 23. Half of the records fall within the period April 10–30. Fourteen final spring sightings in Nebraska are from April 13 to May 25, with a median of May 7. Nineteen initial fall sightings are from July 21 to October 21, with a median of September 30. Thirteen final fall sightings are from October 2 to December 29, with a median of October 24. A state-level analysis of four decade-long periods of Christmas Bird Counts (1967-68 to 2006–7) extending from North Dakota to the Texas panhandle indicated a late-December population peak in Oklahoma (Johnsgard and Shane, 2009).

Habitats: Migrants are found in marshy areas, and during the breeding season nesting occurs in marshes or other wetlands having such vegetation as cattails, phragmites, and shrubs or small trees. In North Dakota the species generally frequents alkaline bogs (fens), especially those having cattails, phragmites, and shrubs or small trees. It also breeds in wet meadows, along swampy shorelines of lakes or streams, and to a limited degree in coastal meadows.

Breeding Status: Breeds southward through eastern South Dakota, and central Nebraska. It is a local summer resident in central Nebraska marshes, including ones in Antelope, Boone, Brown, Garden, Howard, Keith, Loupe, Phelps, Rock, Sheridan and Wheeler counties, and perhaps also nests locally along the Missouri River. There were three confirmed nestings during breeding bird atlas surveys in South Dakota (Peterson, 1995), six in Nebraska (Mollhoff, 2001), and none in Kansas (Busby & Zimmerman, 2001).

Nest Location: Nests are rarely on the ground but instead are built about a foot above the substrate, often in water up to 24 inches deep. They are constructed among the stalks of cattails or in bushes and are rather bulky structures of grass with a finer grass lining.

Clutch Size and Incubation Period: From 3–6 eggs, usually 4–5. The eggs are pale green with dots, spots, and blotches of brown. The incubation period is 12–13 days. Normally single-brooded, sometimes two broods.

Time of Breeding: In North Dakota adults carrying food have been seen as late as August 8. Nests in South Dakota have been reported from June 6 to July 11 (Peterson, 1995).

Breeding Biology: Probably because it is restricted to wet and inaccessible habitats, rather little is known of the breeding biology of this relatively insectivorous sparrow. A breeding density of two pairs in 9 1/2 acres of Maryland bog has been reported, but territory size and other aspects of breeding remain essentially unstudied. Apparently the female incubates alone, although the male has been observed feeding his brooding mate. The nestling period was judged to be about 12–13 days by one early observer, and about nine days by a more recent one, which seems more probable.

Comments: This is a semi-colonial nester whose habitat needs seemingly limit it to only a few known nesting locations in Nebraska. It rather strongly resembles a chipping sparrow, but its trilled song is slower and more melodious. National Breeding Bird Surveys between 1966 and 2009 indicate that the species underwent a statistically non-significant population increase (0.7 percent annually) during that period. Its North American population has been estimated at about nine million birds (Rich *et al.*, 2004).

Suggested Reading: Bent 1968; Rising, 1996; B.O.N.A. 279.

Family Icteridae - Meadowlarks, Blackbirds, Orioles and Allies

Bobolink, *Dolichonyx oryzivorus*

A spring and fall migrant throughout Nebraska, fairly common in central Nebraska, but less common in the eastern and western areas. Breeding occurs from eastern South Dakota south locally to central Kansas.

Migration: The range of 116 initial spring sightings in Nebraska is from March 20 to June 20, with a median of May 16. Half of the fall Nebraska records fall within the period July 29–August 20.

Habitats: While in Nebraska this species is usually found in ungrazed to lightly grazed medium to tallgrass prairies, wet meadows, retired croplands, and occasionally extends to small-grain croplands.

Breeding Status: Breeds in suitable habitats southward through South Dakota (rarely in the Black Hills), most of Nebraska west to Sioux County in the Panhandle, Garden County in the Sandhills, and with the southern limits occurring between the Platte and Republican rivers, and locally or sporadically in northern Kansas (records for Stafford, Cloud, and Barton counties). There were 14 confirmed nestings during breeding bird atlas surveys in South Dakota (Peterson, 1995), 15 in Nebraska (Mollhoff, 2001), and two in Kansas (Busby & Zimmerman, 2001).

Nest Location: Nests are simple hollows scraped into the ground or natural depressions suitable in depth and size, such as that made by a horse's hoof. The nest is invariably concealed in dense vegetation such as tall grasses, clover, or other thickly growing plants. The hollow is filled with grasses and weeds and lined with fine grasses.

Clutch Size and Incubation Period: From 4–7 eggs (four North Dakota clutches averaged 4.8). The eggs are gray to cinnamon, rather irregularly blotched and spotted with brown, and sometimes are almost entirely brown. The incubation period is 13 days. Single-brooded.

Time of Breeding: In North Dakota the breeding season extends from late May to mid-August, with egg dates ranging from June 4–27, and fledglings seen from July 8 to August 16. Nests in South Dakota have been reported from June 6 to July 11 (Peterson, 1995). A Nebraska egg record is from June 14 (Mollhoff, 2001). In Kansas, eggs have been found from June 18 to July 16, young in the nest June 19–29, and recently fledged young from June 30 to July 14.

Breeding Biology: Males arrive on their breeding areas about a week before females and quickly spread out, although specific territorial establishment and defense seems to be weak or lacking. Although the nests are well scattered, males tolerate other males surprisingly near the nest site. The female incubates alone; the male seldom visits her and apparently never feeds her. But males do help feed the young. Broods usually remain in the nest for about 10–14 days but have been reported to leave when only 7–9 days old, or well before they are able to fly. Males often acquire second mates after their first mate has begun nesting. These secondary mates tend to lay smaller clutches than the primary mates, perhaps because they often are young birds or are renesting. This smaller clutch size of secondary mates is adaptive, since males less frequently assist in feeding their second broods, and unassisted females are more likely to be able to tend smaller broods.

Comments: National Breeding Bird Surveys between 1966 and 2004 indicate that the species underwent a statistically significant population decline (2.1 percent annually) during that period. Its North American population has been estimated at about 11 million birds (Rich *et al.,* 2004).

Suggested Reading: Kingsbury 1933; Martin 1971; B.O.N.A. 176; Johnson & Igle, varied dates.

Red-winged Blackbird, *Agelaius phoeniceus*

An abundant spring and fall migrant throughout the region, and a common to abundant summer resident throughout the entire region in suitable habitats.

Migration: The range of 90 initial spring sightings in Nebraska is from January 1 to May 26, with a median of March 3. Half of the records fall within the period February 17–March 17. Eighty final fall sightings range from August 8 to December 31, with a median of November 21. Half of the records fall within the period November 3 to December 21. A state-level analysis of four decade-long periods of Christmas Bird Counts (1967-68 to 2006–7) extending from North Dakota to the Texas panhandle indicated a late-December population peak in Kansas (Johnsgard and Shane, 2009).

Habitats: Breeding occurs on a wide range of habitats, from deep marshes or the emergent zones of lakes and impoundments, through progressively drier habitats such as wet meadows, ditches, brushy patches in prairie, hayfields, and weedy croplands or roadsides. Migrants often are seen in flocks of other blackbird species, feeding in fields or elsewhere, but roosting is typically done in wet areas rather than in residential locations.

Breeding Status: Pandemic, breeding throughout the region in suitable Habitats: There were 164 confirmed nestings during breeding bird atlas surveys in South Dakota (Peterson, 1995), 209 in Nebraska (Mollhoff, 2001), and 533 in Kansas (Busby & Zimmerman, 2001).

Nest Location: Nests are in herbaceous or woody vegetation that is usually in or near water, but they may also be in weeds, bushes, or trees some distance from water, and, rarely, up to 14 feet above the substrate. Of 48 North Dakota nests, hardstem bulrush provided the most common vegetational support, and the rims of 28 nests built over water averaged 13 inches above the water surface. Eight terrestrial nests averaged 19 inches above ground. The nest is built of leaves of grasses and sedges woven together and bound to adjacent vegetation and is lined with fine grasses.

Clutch Size and Incubation Period: From 2–7 eggs (38 North Dakota clutches averaged 3.6, and 243 Oklahoma clutches averaged 3.4). The eggs are pale bluish green, with scrawls and spots of dark tones, mostly toward the larger end. The incubation period is 10–12 days. Frequently double-brooded, and males are often polygynous, with up to six females per male.

Time of Breeding: North Dakota egg dates are from May 15 to July 13, and nestlings are seen from June 9 to July 28. Nests in South Dakota have been reported from May 12 to August 6 (Peterson, 1995). Nebraska egg records are from May 17 to July 8 (Mollhoff, 2001). Kansas egg dates are from March 29 to July 21, with nearly three-fourths of the eggs laid between May 11 and June 10. Young in the nest have been seen there from May 8 to July 21, and recently fledged young from June 9 to July 30, as well as one very late record of September 28.

Breeding Biology: This is one of the commonest and most thoroughly studied of all North American songbirds. Adult males arrive on their breeding marshes well before females and begin to advertise their territories by flight song and "song-spread" displays, both of which prominently exhibit the red upper wing-coverts. Experiments with surgically muting males or painting these red markings black before they acquire mates resulting in the loss of territories by such altered males, although later alteration has no obvious effect. Pair bonds last only during the breeding season, and most territorial males manage to acquire at least two females. In one Wisconsin study, it was found that experienced males tend to return to their old territories in successive years, and that first-year males are usually unable to hold territories long enough to breed. In that study, no more than three females were mated to a single male, but a few instances of double-brooding were found. The young birds leave the nest at 10–11 days, but are dependent for some time thereafter.

Comments: This is one of the region's most abundant breeding birds, numbering in the tens of millions, and also one of the most attractive. Like several other grassland nesting birds it is impacted greatly by brown-headed cowbirds; few nests of redwings in Nebraska seem to survive without being parasitized. National Breeding Bird Surveys between 1966 and 2009 indicate that the species underwent a statistically significant population decline (1.0 percent annually) during that period. Its North American population (north of Mexico) has been estimated at about 200 million birds (Rich *et al.*, 2004).

Suggested Reading: Nero 1956; Peek 1971; B.O.N.A. 184.

Yellow-headed Blackbird, *Xanthocephalus xanthocephalus*

A common to abundant spring and fall migrant throughout the region, and a locally common summer resident in permanent marshes. Breeding occurs from South Dakota south locally to Kansas.

Migration: The range of 103 initial spring sightings in Nebraska is from January 1 to June 5, with a median of April 21. Half of the records fall within the period April 11 to May 1. Eighty-two final fall sightings range from July 23 to December 28, with a median of September 18. Half of the records fall within the period September 4–30. A state-level analysis of four decade-long periods of Christmas Bird Counts (1967-68 to 2006–7) extending from North Dakota to the Texas panhandle indicated a late-December population peak in northwestern Texas (Johnsgard and Shane, 2009).

Habitats: During the breeding season this species occurs in deep marshes, the marsh zones of lakes or shallow impoundments, and elsewhere where there are extensive stands of cattails, bulrushes or phragmites. It is often found breeding in association with red-winged blackbirds, utilizing the deeper portions of the marsh. Migrants are sometimes seen flying or perching with groups of red-winged blackbirds, but more often remain separate from them.

Breeding Status: Breeds in suitable habitats southward from South Dakota (statewide except the Black Hills) through Nebraska, and northwestern Kansas, reaching its normal limits north and west of a line drawn from Meade to Douglas counties, Kansas. There were 64 confirmed nestings during breeding bird atlas surveys in South Dakota (Peterson, 1995), 50 in Nebraska (Mollhoff, 2009), and six in Kansas (Busby & Zimmerman, 2001).

Nest Location: Nests are usually clustered in stands of emergent vegetation, most frequently (in North Dakota) in hardstem bulrush or cattails, with other emergent plants used relatively little. In a sample of 79 nest sites, the water depth ranged from 3–32 inches and averaged 18 inches, and the height of the nest rim above water also averaged 18 inches. By comparison, 28 nests of red-winged blackbirds were in water averaging only nine inches deep, and eight other nests were in terrestrial locations.

Clutch Size and Incubation Period: From 3–7 eggs (109 North Dakota clutches averaged 3.7). The eggs are off-white with spots and dots of browns and grays. The incubation period is 12–13 days. Single–brooded, but males are often bigamous.

Time of Breeding: North Dakota egg dates are from May 10 to July 13, with nestlings seen as early as May 27 and dependent young as late as August 10. Nests in South Dakota have been reported from May 17 to July 18 (Peterson, 1995). Nebraska egg records are from May 28 to June 30 (Mollhoff, 2001). Kansas egg dates are from May 14 to July 5, young in the nest from June 7 to July 9, and recently fledged young from June 13 to July 30..

Breeding Biology: The displays of the yellow-headed blackbird are very similar to those of the red-winged blackbird, but the species differs ecologically in that the males normally participate in brood care, are more dependent on emerging aquatic insects such as damselflies, and thus are more dependent on marshes than are redwings. In both species, the males' conspicuous and prolonged displays seem to be related to the importance of territorial size and quality in attracting the maximum number of females. As in the red-winged blackbird, only the female incubates, but males often help feed the young, particularly those of their first female. The young leave the nest at 9–12 days.

Comments: This very attractive species seems to prefer somewhat alkaline marshes more than freshwater ones, and thus it becomes more common in the Sandhills marshes as one proceeds westward. National Breeding Bird Surveys between 1966 and 2009 indicate that the species underwent a statistically non-significant population decline (0.6 percent annually) during that period. Its North American population has been estimated at about 23 million birds (Rich *et al.,* 2004).

Suggested Reading: Willson 1964; Orians and Christman 1968; B.O.N.A. 192.

Non-breeding Species of Wetland Birds

(62 spp.)

ORDER ANSERIFORMES - WATERFOWL

Family Anatidae - Swans, Geese and Ducks

Greater White-fronted Goose, *Anser albifrons*

A common spring and fall migrant throughout the region, being most abundant in Nebraska's central Platte Valley and Rainwater Basin. Wintering occurs in the southern parts of the region. An estimated 350,000 were seen at Quivira N.W.R., during the Audubon Christmas Count of 2004–5.

Migration: Twenty-nine initial spring sightings in Nebraska are from February 12 to May 12, with a median of March 12. Seventeen final spring sightings in Nebraska are from March 23 to May 18, with a median of April 14. Nineteen initial fall sightings are from September 14 to November 21, with a median of October 23. Fifteen final fall sightings are from October 12 to December 29, with a median of November 6. A state-level analysis of four decade-long periods of Christmas Bird Counts (1967-68 to 2006–7) extending from North Dakota to the Texas panhandle indicated a late-December population peak in Kansas (Johnsgard and Shane, 2009).

Habitats: Migrants are associated with large marshes, shallow lakes, wide rivers with bars and islands, and adjacent agricultural grain fields.

Comments: This fine goose concentrates in the Platte Valley in spring, with about 80 percent of the entire mid-continent population are then concentrated in the state. It breeds in the high Canadian arctic, often near snow geese. The 2009 mid-continent fall population of this goose was about 752,000, well below many earlier estimates (U.S.F.W.S., 2009a), and per-

sonal observations in the Platte valley of Nebraska suggest there has been a very substantial decline in spring numbers there.

Suggested Reading: Johnsgard, 1975; B.O.N.A. 131; Kear, 2005.

Snow Goose, *Chen caerulescens*

A spring and fall migrant throughout the three-state region. Less common westward but abundant in the Missouri River Valley and the central Platte Valley. The bluish-colored morph ("blue goose") and less common intermediate (heterozygotic) plumage types comprise about 20–25 percent of the total population in eastern parts of the region, but are less frequent westward. Migrants are abundant throughout the Great Plains during spring and fall, particularly in the Missouri River Valley, which supports a mid-continental population of more than two million birds.

Migration: Thirty-six initial spring sightings in Nebraska range from January 8 to March 28, with a median of March 9. Twenty-six final spring sightings in Nebraska are from March 6 to May 20, with a median of April 20. Forty initial fall sightings are from August 19 to December 16, with a median of October 4. Thirty-eight final fall sightings are from October 26 to December 31, with a median of December 2. Wintering occurs in the southern parts of the region; 345,000 were seen at Quivira N.W.R., during the Audubon Christmas counts of 2004–5. During the 2008–2009 count, Quivira N.W.R. held 140,000 snow geese, of which 21 percent were of the blue morph plumage type. Over the four-decade period 1967–2006, blue-morph birds made up 27.5 percent of all the snow geese counted in the five-state Plains region. This proportion declines rapidly one moves westward across the Great Plains. Over time the proportion of blue-morph birds in the central and western states has slowly increased, for reasons that are still uncertain. A state-level analysis of four decade-long periods of Christmas Bird Counts (1967-68 to 2006–7) extending from North Dakota to the Texas panhandle indicated a late-December population peak in Kansas (Johnsgard and Shane, 2009). The large numbers reported on recent Kansas counts are probably a reflection of late autumn migrations rather than of wintering birds. However, certainly far more geese (especially snows and Canadas) now winter along the Missouri Valley of Kansas and Missouri than was the case during the 1960's.

Habitats: Marshes, sloughs, riverbottom meadows and croplands such as cornfields are used on Migration: Lakes or reservoirs near croplands are also utilized.

Comments: Snow geese in the Great Plains have increased tremendously in the past few decades; current populations of five or more million birds are more than their tundra breeding grounds can support. The historic Missouri Valley migrant flock has also shifted its migration route somewhat farther west into the Platte Valley of central Nebraska, especially during spring. Up to a million snow geese now sometimes occur on large Rainwater Basin marshes, such as Harvard and Funk waterfowl production areas. This migration shift has brought snow geese into competitive contact with millions of migrating cackling, Canada and greater white-fronted geese, and a half-million sandhill cranes. Winter or spring 2009 national population estimates included nearly four million lesser snow geese (U.S.F.W.S., 2009a). All the populations of this species were still increasing as of 2009 (Johnsgard, 2010). The estimated average U.S. kill of snow geese during the 2004–8 seasons was 565,000, of which about 27 percent were blue-morph. Estimated total annual Canadian kills from 1990–1998 ranged from about 38,000–106,000 for white-morph lessers, and 33,000–66,000 for blue-morph lessers. The annual Canadian kills for greater snow geese during that period ranged from 29,000–102,000. These numbers represent less than ten percent of the estimated continental population of perhaps six million snow geese. They have failed to slow the rate of popu-

lation growth, in spite of a decade of federal efforts to promote almost un-
limited recreational hunting. Spring hunting is highly detrimental to many
other non-target species, which are severely stressed and frightened out of
shared wetland habitats.

Suggested Reading: Johnsgard, 1975, 2010; B.O.N.A. 514; Kear, 2005.

Ross's Goose, *Chen rossii*

An increasingly common and regular spring and fall migrant in the region. It is present each spring in Nebraska's Rainwater Basin wetlands, and
many eastern and Platte Valley counties, typically in the company of large
flocks of snow geese. This goose is now probably a regular and uncommon
migrant throughout the Great Plains, but is often hard to find among flocks
of snow geese.

Migration: Six spring Nebraska records are from March 10 to April 13,
with a mean of March 29. Five fall Nebraska records are from November
10 to December 22, with a mean of November 26. Ross's geese comprised
an estimated two percent of the carcasses among 1,200 white geese killed
by a tornado passing near York, on March 13, 1990 (*Nebraskaland* 68(2):
34–41). As many as 5,600 Ross's geese have been seen at a Nebraska single location containing 14,000 white geese (*Nebraska Bird Review* 69:50),
and during the 2008–2009 count Audubon Christmas Bird Counts Quivira
N.W.R. held over 29,000 Ross's geese. A state-level analysis of four decade-
long periods of Christmas Bird Counts (1967-68 to 2006–7) extending from
North Dakota to the Texas panhandle indicated a late-December population
peak in Oklahoma (Johnsgard and Shane, 2009).

Migration: Spring migration is early; in Nebraska the peak is about
March 15, as in snow geese. Fall migration is also probably comparable to
that of snow geese.

Habitats: Found in the same habitats as snow geese.

Comments: These tiny geese are easily overlooked among the vast flocks
of snow geese with which they associate. A few blue-morph Ross's geese
have been seen in Nebraska (*Nebraska Bird Review* 66:19; 69:50), and
at least one reported from South Dakota (Tallman, Swanson and Palmer,
2002), but the frequency of this plumage variant is still only one in many
thousands. The genes for this plumage variant probably entered the Ross'
goose's gene pool via hybridization with blue-morph snow geese. Apparent
hybrids between these species have been reported in the region. Because of
difficulties in field separation of Ross's geese from snow geese, no attempts
are made to identify and inventory Ross's geese nationally, but one enor-

mous nesting colony in the tundra lowlands of arctic Canada's Queen Maud Gulf had 726,000 birds in 2008, and comprised a substantial percentage of this species' gradually increasing population (U.S.F.W.S., 2009a). Estimated total annual Canadian kills from 1990–1998 ranged from about 2,000–29,000. As a result of the kill-as-many-white geese-as-possible campaign of the U.S. Fish & Wildlife Service Ross's goose kills in the U.S. have also increased greatly in recent years, with an estimated kill of 106,000 in 2001, and a mean of 78,000 during the five years 2004–8.

Suggested Reading: Johnsgard, 1975, 2010; B.O.N.A. 162; Kear, 2005.

Cackling Goose, *Branta hutchinsii*

A common to occasional spring (early March to mid-April) and fall (early October to early December) migrant throughout the region. Four small tundra-breeding forms (*hutchinsii, taverneri, leucopareia* and *minima*) were designated by the American Ornithologists' Union in 2004 as specifically distinct from the larger and generally more southerly-breeding races of Canada goose. These small, high-arctic breeding, geese now bear the collective English name cackling goose, a name one reserved for the very small dark-bodied geese of the Pacific Coast. Difficulties in visual separation from small races (especially *parvipes*) of the Canada goose make seasonal records and relative abundance estimates somewhat suspect.

Migration: Spring migration is early; in Nebraska the peak is about March 15. Fall migration is less well known, but is probably similar to that of snow geese. Cackling geese and small races of Canada geese are abundant migrants through the region, and up to 100,000 may winter in ice-free locations during mild winters.

Habitats: Found in the same habitats as Canada geese.

Comments: Swenk (*Nebraska Bird Review* 2:103–116, 1934) classified 17 of 404 Nebraska-shot Canada geese as representing the then-subspecies *hutchinsii*. He classified the majority of the birds as the intermediate-sized *leucopareia,* which would probably include the forms represented by the present-day taxa *parvipes* and *taverneri* (the latter is now considered part of *hutchinsii*). Typical pale-plumaged *hutchinsii* types (often called Richardson's geese) are seemingly most common during spring and fall in the central Platte Valley. Recent revisions of the white-cheeked goose complex by H. C. Hansen and B. W. Anderson have recommended that several new taxa be recognized, but their efforts have seemingly only added to the current complexity and confusion surrounding these morphologically diverse birds (*Wilson J. of Ornithology* 119:514–515; 121:658-660; *Auk* 128:805–807). The westernmost of the populations currently considered as cackling geese is the Aleutian cackling goose (*leucopareia*). It was once listed as nationally endangered and limited to the outermost Aleutians, but surged as a result of hunting restrictions, transplants to new breeding islands and effective predator control, reaching 100,000 birds by 2009. The 2009 estimates (U.S.F.W.S., 2009a) for the other cackling goose races included 160,000 for nominate *minima* of the Pacific coast, 60,000 for the Alaska cackling goose *taverneri*, and 220,000 for the Baffin Island or Richardson's cackling goose *hutchinsii*, the last of which is the common Great Plains migrant. A three-year analysis of Audubon Christmas Bird Counts (2004-05 to 2006–7) extending from North Dakota to the Texas panhandle indicated a late-December population peak in Kansas (Johnsgard and Shane, 2009). As many as

12,000–14,000 have been observed in the Lake McConaughy area in January (Brown, Dinsmore, & Jorgensen, 2011).

Suggested Reading: Kear, 2005; Silcock, 2006.

Tundra Swan, *Cygnus columbianus*

An uncommon or occasional spring and fall migrant in South Dakota. Less common in the rest of the three-state region, and very rare in Kansas, mainly in the eastern parts of that state.

Migration: Twenty spring Nebraska sightings range from January 1 to May 15, with a median of March 27. Eleven fall sightings are from October 21 to December 14, with a median of November 22.

Habitats: Shallow lakes, marshes and adjacent flooded fields are used by migrants.

Comments: Previously known as the whistling swan, this species' current name reflects a taxonomic merger with the Bewick's swan of the Old World, and describes its tundra breeding habitat very well. Most tundra swans miss the central Great Plains while migrating, by either turning east in Minnesota and eventually reaching the Chesapeake Bay region of the Atlantic coast, or veering west in North Dakota, heading toward the Great Salt Lake region of Utah and wintering grounds in the Central Valley of California. During the late 1980s the overall North American estimates were of about 87,000 tundra swans in the western population, which has been "experimentally" hunted since 1962, and about 64,000 in the eastern one, which has been hunted since 1984 (Kear, 2005). By 2009 the two populations were estimated to total about 100,000 birds each (U.S.F.W.S., 2009a).

Suggested Reading: Johnsgard, 1975; B.O.N.A. 89; Kear, 2005.

Eurasian Wigeon, *Anas penelope*

An occasional migrant through the region, at least during spring, when plumage differences from the American wigeon are most evident.

Migration: Most records are for March and April, the latest is May 2. In recent years several reports suggest the species to be a regular spring migrant, especially in the Rainwater Basin. Nebraska county records include at least Adams, Cedar, Clay, Cuming, Dawes, Garden, Lincoln, Seward, and Thayer counties One probable hybrid wigeon has also been reported (*Nebraska Bird Review* 66:35). There were nine accepted Kansas records since 1950 (Thompson *et al.*, 2011), and at least three are known from South Dakota (Tallman, Swanson and Palmer, 2001).

Habitats: Usually found among flocks of American wigeons in the three-state region, and using the same habitats during migration.

Comments: During recent hunting seasons a maximum of 190 Eurasian wigeons were killed in the Atlantic flyway, and a maximum of 2,120 in the Pacific flyway, but relatively few are reported in the interior parts of the continent. Total U.S. hunter-kills have averaged about 1,200 annually since 1994; Eurasian wigeons were apparently not distinguished from American wigeons during earlier U.S. hunter-kill surveys. Estimated total annual Canadian kills from 1990–1998 ranged from about 50–300. Thus, Eurasian wigeons comprised about 0.3–1.5 percent of all wigeons identified in the U.S. and Canadian kills, suggesting that the North American population numbers in the hundreds of thousands. In spite of all these occurrences, there is still no evidence of Eurasian wigeons breeding in North America, which might be occurring in remote parts of coastal Alaska.

Suggested Reading: Johnsgard, 1975; Kear, 2005.

Greater Scaup, *Aythya marila*

An occasional or uncommon migrant and winter visitor the three-state region, probably more common than the available records would suggest. It is probably regular during late fall, winter, and early spring on larger reservoirs and lakes. Although migrants might appear anywhere in the Plains States, they are probably most common in the Missouri Valley impoundments.

Migration: Twenty-seven total spring Nebraska records are from January 11 to May 18, with the largest number (12) for March, followed by April (8), and three each for February and May. There are fall Nebraska records from October 27 to December 30.

Habitats: Migrants and wintering birds utilize lakes and reservoirs in the interior, but most birds winter coastally.

Comments: Males of the two scaup species are quite similar (and the females even more), but greater scaups have a greenish-glossed head (not purplish), and a much flatter crown profile, with no hint of a crest at the rear. The two species don't often associate because of their differing habitat preferences. A North American population about 400,000 seems possible, based on surveys of both scaup species collectively. The average annual hunter-kill estimate in the U.S. during the five years 2004–8 has been about 59,000 birds, but averages have been in a long-term decline since the 1960's. Estimated total annual Canadian kills from 1990–1998 ranged from about 12,000–27,000.

Suggested Reading: Johnsgard, 1975; B.O.N.A. 650; Kear, 2005.

Surf Scoter, *Melanitta perspicillata*

A rare migrant the three-state region, occurring primarily in the fall. Like the other scoters, most individuals are females or immature males that are easily overlooked or confused with other species.

Migration: Five spring Nebraska records are from April 21 to May 15. Eight fall Nebraska records are from October 7 to December 16, with a mean of November 6. Sharpe, Silcock and Jorgensen (2001) noted eight spring reports, from April 5 to May 15, and 30-35 fall reports, from October 5 to December 20.

Habitats: Lakes, reservoirs and larger rivers are used by migrants. Most wintering is done coastally.

Comments: About as rare as the preceding species, birds seen in Nebraska have a distinctive head pattern, with whitish spots in front of and behind the eye, and lack white wing markings. This species' population is still only very poorly documented, but Rose & Scott (1997) suggested a stable population of 765,000 birds. The average annual hunter-kill estimate in the U.S. during the five years 2004–8 has been about 33,400 birds, and estimates have exhibited a gradually increasing long-term trend since the 1960's. Estimated total annual Canadian kills from 1990–1998 ranged from about 5,000–19,000.

Suggested Reading: Johnsgard, 1975; B.O.N.A. 363; Kear, 2005.

White-winged Scoter, *Melanitta fusca*

An occasional spring and fall migrant in the region., more common in the fall than spring, and with most of the Nebraska records from counties bordering the Platte or Missouri rivers. It is probably an annual visitor to the region, especially along the Missouri River and on the larger reservoirs.

Migration: Five total spring Nebraska records are from March 31 to April 29, with a mean of April 7. Twenty-one fall Nebraska records are from October 7 to December 10, with a median of November 10. Half of the records fall within the period October 28 to November 22. Sharpe, Silcock and Jorgensen (2001) noted 17 spring reports, from March 26 to May 4, and about 55 fall reports, from October 15 to December 20. It was observed five years during Christmas Bird Counts at Lake McConaughy between 2000–2001 and 2011–2012.

Habitats: Lakes, reservoirs and larger rivers are used by some migrants. Most birds migrate and winter in coastal areas.

Comments: This largest of the scoters is similar to the surf scoter in female or immature plumage, but has a large white patch on the inner wing feathers that allow for positive identification. It is the species most often seen in Nebraska, and once bred as close as northern North Dakota. The North American population of this species may consist of about 1,000,000 birds (Rose & Scott, 1997; Kear, 2005). The average annual hunter-kill estimate in the U.S. during the five years 2004–8 has been about 8,500 birds, but estimates have exhibited a gradually declining long-term trend-line since the 1960's. Estimated total annual Canadian kills from 1990–1998 ranged from about 4,000–10,000.

Suggested Reading: Johnsgard, 1975; B.O.N.A. 274; Kear, 2005.

Black Scoter, *Melanitta nigra*

An extremely rare migrant in three three-state region, occurring primarily in fall. The rarest of the regional scoters, it is rare or accidental in South Dakota and Kansas, and has been seen in a few scattered locations throughout Nebraska.

Migration: Two spring Nebraska records are for March 25 and May 4. Seven fall Nebraska records range from September 28 to December 10, with a mean of October 28. Sharpe, Silcock and Jorgensen (2001) noted seven spring reports, from February 25 to May 25, and about 35 fall reports, from October 21 to late December, plus a winter (January 16) record.

Habitats: Lakes, reservoirs and larger rivers are used by migrants. Most wintering is done coastally.

Comments: This rare sea duck usually is seen in the female-like immature plumage while in the three-state region, when its two-tone brown head helps in identification. Its North American population (sometimes considered as a species separate from the Eurasian population) is still unknown The average annual hunter-kill estimate in the U.S. during the five years 2004–8 has been about 12,000 birds, but yearly estimates have been rather variable. Estimated total annual Canadian kills for seven years between 1969 and 1993 ranged from about 4,000–8,000.

Suggested Reading: Johnsgard, 1975; B.O.N.A. 177; Kear, 2005.

Long-tailed Duck, *Clangula hyemalis*

A rare fall and spring migrant throughout the region, but perhaps slightly more common eastward. Two or more records exist for Keith, Lancaster, Douglas and Washington counties of Nebraska, with at least eight for Douglas County. Increasingly common since the formation of large regional reservoirs. As of 1933 there were less than a dozen definite records for the state (*Nebraska Bird Review* 1:11), but the species was seen almost every year in the 1970s, and is now regular during early winter at Lake Mc-Conaughy and Lake Ogallala, with up to nine having been seen at one time in early December.

Migration: Thirteen total spring Nebraska records are from February 3 to April 19, with a median of March 29. Ten total fall Nebraska records are from October (no date) to December 11, with a median of November 27. Sharpe, Silcock and Jorgensen (2001) noted spring reports as extending from March 2 to May 18, and fall reports from October 31 to late December, plus 13 winter reports.

Habitats: Lakes, reservoirs and larger rivers are used by migrating birds; most wintering is done in coastal habitats.

Comments: Long-tailed ducks are circumpolar and nest higher in the arc-tic than almost any other duck, and mostly occur in the three-state region as late fall and winter vagrants. The world population may include about 2.7 million birds in North America (Rose & Scott, 1997), making it by far the most abundant of our sea ducks. The average annual hunter-kill esti-mate in the U.S. during the five years 2004–8 has been about 28,200, and estimates have exhibited a long-term progressive increase since the 1960's. Estimated total annual Canadian kills from 1990–1998 ranged from about 5,000–10,000. The very small estimated hunter-kill relative to its huge con-tinental population is probably a reflection of this species' mostly marine, high-latitude distribution and its relatively low attractiveness to hunters.

Suggested Reading: Johnsgard, 1975; B.O.N.A. 651; Kear, 2005.

Common Goldeneye, *Bucephala clangula*

A common to uncommon spring and fall migrant throughout the region, occasionally over-wintering where open water is available. Breeding occurs in Minnesota and locally in North Dakota.

Migration: Thirty-five initial spring sightings in Nebraska range from January 1 to April 12, with a median of March 5. Twenty-four final spring sightings in Nebraska are from March 9 to May 8, with a median of March 30. Thirty-four initial fall sightings are from October 10 to December 31, with a median of November 21. Thirty-one final fall sightings are from No-vember 22 to December 31, with a median of December 14. A state-level analysis of four decade-long periods of Christmas Bird Counts (1967-68 to 2006–7) extending from North Dakota to the Texas panhandle indicated a late-December population peak in Kansas (Johnsgard and Shane, 2009).

Habitats: Deeper marshes, rivers, lakes and reservoirs are used during migration. Large unfrozen lakes and reservoirs are important wintering areas.

Comments: This beautiful diving duck appears in spring at about the same time as do common mergansers, and both species can often be seen engaging in excited courtship displays, performing head-throws, backward kicks, vertical neck-stretching, and other remarkable male posturing, as the drab females either ignore the males or perhaps threaten them with bill-pointing gestures. The white plumages of male goldeneyes and common mergansers fairly glisten in the sun, and when in flight both species exhibit large while wing-patches. The entire North American population of this northern hemisphere species might total about 1.5 million birds (Kear, 2005). The average annual hunter-kill estimate in the U.S. during the five years 2004–8 has been about 75,400 birds, and estimates have been quite stable since the 1960's. Estimated total annual Canadian kills from 1990–1998 ranged from about 25,000–77,000.

Suggested Reading: Johnsgard, 1975; B.O.N.A. 170; Kear, 2005.

Barrow's Goldeneye, *Bucephala islandica*

A rare winter and spring migrant in the region, probably mainly occurring in the west. It is regular but rare during winter in Pennington County, South Dakota, and is very rare in Kansas. In Nebraska, it has been observed two or more times in several counties. By the late 1990's there were at least ten records for Lake McConaughy (*Nebraska Bird Review* 66:11), and it was observed three years during Christmas Bird Counts at Lake McConaughy between 2000–2001 and 2009–2010.

Migration: Eight spring Nebraska records range from February 15 to April 2, with a mean of March 19. Three fall Nebraska records are from November 26 to December 21. Sharpe, Silcock and Jorgensen (2001) noted eight spring reports, from March 3 to April 19, and nine fall reports, from October 30 to late December, and many winter records extending into February. A few birds winter as far north as South Dakota.

Habitats: While on migration this species uses the same habitats as the common goldeneye, but is more prone to winter in coastal or brackish waters.

Comments: Although it breeds fairly commonly in Wyoming's mountain ranges, there is little chance of seeing this species in the region, and recognizing females is difficult. A few hybrids with common goldeneyes have been recorded, but such hybrids are even more difficult to recognize. Minimum population estimates exist for Alaska (45,000), British Columbia (70,000–126,000) and the Pacific Coast states (under 8,000) (Kear, 2005). Small numbers also occur along the Rocky Mountain range south locally to Wyoming and northern Colorado, from which the Great Plains birds probably originate. The average annual hunter-kill estimate in the U.S. during the five years 2004–8 has been about 5,200 birds, and estimates have remained fairly stable since the 1960's. Estimated total annual Canadian kills from 1990–1998 ranged from about 500–3,700.

Suggested Reading: Johnsgard, 1975; B.O.N.A. 548; Kear, 2005.

Red-breasted Merganser, *Mergus serrator*

An occasional to rare spring and fall migrant throughout the region. It perhaps over-winters rarely, and has most often been seen at Lake McConaughy and other large reservoirs.

Migration: Sixty-one initial spring sightings in Nebraska are from January 15 to May 12, with a median of March 29. Half of the records fall within the period March 19 to April 7. Twenty-four final spring sightings in Nebraska are from February 14 to May 18, with a median of April 20. Sixteen total fall sightings are from September 21 to December 31, with a median of November 18. Half of the records fall within the period November four through November 27. A state-level analysis of four decade-long periods of Christmas Bird Counts (1967-68 to 2006–7) extending from North Dakota to the Texas panhandle indicated a late-December population peak in Oklahoma (Johnsgard and Shane, 2009).

Habitats: Lakes, reservoirs and large rivers are used by migrants; wintering more often occurs coastally.

Comments: The red-breasted merganser is less common in the region than the two other mergansers, which is unfortunate because it is a most attractive species. Like the common merganser, it is dependent upon fish in clear-water habitats. World population estimates of this widely ranging (Holarctic) species include 237,000 birds in North America (Rose & Scott, 1997; Kear, 2005). The average annual hunter-kill estimate in the U.S. during the five years 2004–8 has been about 15,000 birds, and estimates appear to have remained fairly stable since the 1960's. Estimated total annual Canadian kills from 1990–1998 ranged from about 7,000–16,000.

Suggested Reading: Johnsgard, 1975; B.O.N.A. 443; Kear, 2005.

ORDER GAVIIFORMES - LOONS

Family Gaviidae - Loons

Red-throated Loon, *Gavia stellata*

A very rare spring and fall migrant throughout the three-state region. It has been reported from 22 counties in Kansas (Thompson, 2011), and at least five times in South Dakota (Tallman, Swanson and Palmer, 2002). Many Nebraska records are for Douglas and Lancaster counties, but it has also been observed in Buffalo, Frontier, Keith, Sarpy and Washington counties,

Migration: Five spring Nebraska records are from April 17 to May 7, with a mean of April 28, and eight fall Nebraska records are from October 31 to December 2, with a mean of November 17. There is also one mid-June record. Sharpe, Silcock and Jorgensen (2001) listed 10 spring reports, from April 6 to May 16, and 8 documented fall reports, from October 30 to November 25, nearly all in the east.

Habitats: Larger rivers, lakes and reservoirs are used while on migration.

Comments: This is the smallest of the loons, and the only one lacking the black-and-white back patterning when in breeding plumage. It is also the only loon that can take off from small ponds, which are often common on its arctic nesting grounds. There are no available world population estimates of this circumpolar species.

Suggested Reading: Johnsgard, 1987; B.O.N.A. 513.

Pacific Loon, *Gavia pacifica*

A very rare migrant throughout the three-state region. It has been reported from 23 counties in Kansas (Thompson, 2011), and at least six times in South Dakota (Tallman, Swanson and Palmer, 2002), and nearly 40 Nebraska records for this species as of 2001 (Sharpe, Silcock and Jorgensen, 2001).

Migration: Many of the Nebraska records are for November; there is one spring record. Sharpe, Silcock and Jorgensen (2001) listed two spring reports, from April 3 to May 16, five summer records, and about 30 fall reports, mostly for the period November 2–24.

Habitats: Larger rivers, lakes and reservoirs are used while on migration.

Comments: This is and intermediate-sized loon that was once part of the species known as the "black-throated loon" before being split into two very similar species; the very similar Arctic loon replaces the Pacific loon in Eurasia. There are no available world population estimates of this high-arctic breeding species.

Suggested Reading: Johnsgard, 1987; B.O.N.A. 657.

Common Loon, *Gavia immer*

Uncommon spring and fall migrant throughout the three-state region. This species has been observed in at least 33 Nebraska counties, being recorded ten or more times in Douglas, Lincoln and Lancaster counties, at least five times in Scotts Bluff and Keith counties. It is uncommon in South Dakota, with some birds remaining through the summer, but with no proof of breeding. It has been seen in the majority of Kansas counties. Most of the sightings have occurred in the eastern half of the three-state region.

Migration: A total of 55 initial spring sightings in Nebraska range from March 18 to May 27, with a median of May 7. Fourteen final spring sightings in Nebraska are from April 12 to May 28, with a median of May 16. Twenty-five initial fall sightings are from July 20 to November 2, with a median of October 24. Seventeen final fall sightings are from October 25 to December 7 with a median of November 2. Of a total of 135 records, the largest number (37) are for April, followed by May (35), November (26), and October (15). Records exist for all months. A state-level analysis of four decade-long periods of Christmas Bird Counts (1967-68 to 2006–7) extending from North

Dakota to the Texas panhandle indicated a late-December population peak in Oklahoma (Johnsgard and Shane, 2009).

Habitats: Larger rivers, lakes and reservoirs are used while on migration.

Comments. This is by far the most common of the loons and has been seen in Nebraska during all twelve months. Many of the birds seen are in immature or winter plumage, but some breeding-plumaged birds may be seen during spring migration. There are no available world population estimates of this Holarctic species. Johnsgard (1987) estimated that the breeding population south of Canada might number about 15,000 birds, but the great majority of North American loons breed in Canada and Alaska.

Suggested Reading: Johnsgard, 1987; McIntyre, 1988; B.O.N.A. 313.

ORDER PELECANIFORMES – PELICANS & CORMORANTS

Family Threskiornithidae–Ibises and Spoonbills

Glossy Ibis, *Plegadis falcinellus*

Rare in Nebraska and Kansas, with no definite regional nesting records, and am accidental vagrant in South Dakota. About 45 Nebraska records had been accepted for this species as of 2010. Of 33 Nebraska records obtained though 2009, 30 were for the period 2005–2009 (*Nebraska Bird Review* 77:52; 78:44). Sharpe, Silcock and Jorgensen (2001) noted 12 total state reports, nearly all undocumented. There are also 20 records from seven Kansas counties (Thompson *et al.,* 2011,) and from one South Dakota county (Tallman, Swanson and Palmer, 2002).

Migration: Seasonal Nebraska records extend from April 24 to October 5. Kansas records are from April 7 to June 10, and from July 4 to September 13 (Thompson *et al.,* 2011). Assumed to be a migrant only in the region, the species' great similarity to the white-faced ibis makes breeding documentation unusually difficult. Various immature and non-breeding examples of this species have been reported for Nebraska, and two may have nested with white-faced ibises in 2008 (*Nebraska Bird Review* 78:90).

Comments: National Breeding Bird Surveys between 1966 and 2009 indicate that the species underwent a statistically significant population increase (2.9 percent annually) during that period.

Suggested Reading: B.O.N.A. 545.

ORDER GRUIFORMES – CRANES, RAILS & GALLINULES

Family Rallidae–Rails, Coots and Gallinules

Yellow Rail, *Coturnicops noveboracensis*

Apparently an extremely rare spring and fall migrant in the three-state region, but a possible if still unproven breeder in South Dakota. Nesting information is provided here on the assumption that regional nesting might eventually be observed. Most of the Nebraska and South Dakota records are from eastern counties.

Migration: Eight total spring Nebraska records are from April 26 to June 10, with a mean of May 6. There apparently are at least eight fall Nebraska records for the species in Nebraska, from September 16 to October 1 (Sharpe, Silcock and Jorgensen, 2001). It has been seen from May 11 to August 26 in South Dakota (Tallman, Swanson and Palmer, 2002). There are 13 county records from Kansas, from March 5 to May 28 and from late August to October 28 (Thompson *et al.*, 2011).

Habitats: During migration this species is likely to be found in marshes with extensive grassy or sedge vegetation. When they occur in the same marshes with Virginia and sora rails they tend to occupy the densest areas of sedges, while the other species are more often found in areas of cattails and bulrushes.

Breeding Status: A local and elusive species. There are at least four June records for Nebraska, suggesting possible breeding. It also may breed in South Dakota, where there are June and August records for Bennett, Brule and Minnehaha counties (Tallman *et al*, 2002). However, the nearest known breeding records are from North Dakota.

Habitats: In North Dakota, yellow rails are limited to fen-like areas or boggy swales associated with springs. Often they are quaking surface mats of emergent vegetation such as cattails, bulrushes, sedges, and associated species. Yellow rails sometimes nest in the same marshes as sora and Virginia rails but occupy the densest areas of sedges, while the other species occupy cattails and bulrushes.

Nest Location: Nests may be built over wet ground or over water up to four inches deep, usually in dense emergent vegetation consisting of grasses and grass-like plants. The nest is usually under a canopy of dead grass and fairly close to a spring-fed brook. It is a coiled cup of dead grass lined with bits of grasses, sedges, and mosses.

Clutch Size and Incubation Period: From eight to ten eggs, often nine. The eggs are buffy to pinkish with numerous small brown spots. The incubation period is 16–18 days, beginning with the last egg.

Time of Breeding: Egg dates in North Dakota are from May 25 to June 19, with most egg-laying probably occurring in the first ten days of June, and hatched eggs have been seen as early as June 16.

Breeding Biology: During spring, males establish territories in dense marshes, patrolling them frequently and uttering their distinctive clicking notes *(tic-tic, tic-tie-tic)*. Males are immediately evicted, but females are approached with a wing-spreading display. After pair bonds are formed, the mates preen each other and copulations are frequent. Nest-building begins nearly a month before incubation, with both sexes participating, and several extra brood nests are often constructed. The female does the final lining of the nest and apparently performs all the incubation, leaving the nest only to feed for brief periods. She also finishes building a brood nest after the clutch hatches; the chicks hatch nearly synchronously. The female broods and feeds her young both in and out of the nest for about 17 days, after which they are brooded only at night. They are nearly independent by their third week of life but do not fledge until they are about 35 days old. Evidently the male plays no active role in defending or caring for the brood.

Comments: Yellow rails are among the most elusive of birds, and even the most avid birders often fail to add this species to their life lists. Most have to settle for hearing responses to playbacks of the species' calls.

Suggested Reading: Terrill 1943; Cink, 1973; Stalheim 1975; Burt, 1994; B.O.N.A. 139.

Purple Gallinule, *Porphyrio martenica*

This southern species of gallinule is an occasional summer visitor to the eastern half of Kansas, where it has not yet been proven to breed (nesting information is provided here on the assumption that it might eventually be observed). There are 11 county records, including Chase, Coffee, Douglas, Harvey, Johnson, Lynn, Lyon, Montgomery, Riley, Sedgwick and Wyandotte (Thompson *et al.,* 2011). There is a single South Dakota record, from Clark County, May 22, 1999, and none from Nebraska.

Migration: With one exception, Kansas records range from April 4 to June 17. Wintering occurs from Florida and the Gulf Coast south to Central and South America, where the birds are year-around residents.

Habitats: Wetlands with a combination of emergent and floating vegetation are preferred.

Breeding Status: An occasional summer visitor in eastern Kansas, where breeding has not yet been documented.

Habitats: In our region this species is limited to marshes with extensive growth of water lilies, lotus, and other aquatic vegetation. It also occurs in tropical swamps, rivers, lagoons, rice plantations, and similar habitats through much of the western hemisphere.

Nest Location: Nests are built over relatively deep water, sometimes on floating islands, in pickerelweed, or in woody vegetation. The nest is well concealed from above by arched-over vegetation and has a flat lateral runway leading downward to the water.

Clutch Size and Incubation Period: From 5–10 eggs, usually 6–8. The eggs are pale buff with a few small brown spots. The incubation period is 20–23 days, starting before the clutch is completed. Probably double-brooded in southern regions.

Time of Breeding: In Oklahoma, eggs have been seen from May 15 to July 18, and young recorded from May 15 to August 18.

Breeding Biology: Few studies have been done on the behavior and biology of this species, but it resembles the common gallinule in being highly territorial and in advertising courtship and feeding territories with repeated *kuk* or *keek* notes. Nest-building begins a few weeks after the birds arrive and establish territories; the male probably does most of the nest-building. Both sexes incubate, and mates perform a nest-relief ceremony of presenting a leaf to the incubating bird, which incorporates it into the nest before departing. The young hatch over a period of about four days and are brooded actively for about a week, after which they are brooded only at night. One or more brood nests is typically present. The fledging period is uncertain but is probably about 6–7 weeks, and molting adults also apparently undergo a flightless period during late summer.

Comments: Given the influence of global warming, it is possible that nesting by this tropically oriented species might eventually occur in Kansas.

Suggested Reading: Meanley 1963; Trautman and Glines 1964; Tacha and Braun, 1994; B.O.N.A. 626.

Family Gruidae–Cranes

Whooping Crane, *Grus americana*

An occasional spring and fall migrant in the region, more often seen in spring than in fall. It has been observed in at least 26 Nebraska counties, but most commonly in Buffalo and Kearney counties. Over 90 percent of the sightings through the early 1970's have occurred within 30 miles of the Platte River, and about 80 percent have occurred between Lexington and Grand Island (*Nebraska Bird Review* 45:54–6). Sightings in the Platte Valley have greatly increased in recent years as the species' population has increased (Gill & Johnsgard, 2010). It is also rare to occasional in Kansas and South Dakota; Cheyenne Bottoms Wildlife Area and Quivira N.W.R. are its most commonly visited Kansas localities; both Quivira N.W.R. and the central Platte Valley are federally classified as representing critical habitat.

Migration: A summary of Nebraska migration records for this species (*Nebraska Bird Review* 45:54–6), indicates that the spring migration extends from early March to late May, with a peak during the period April 1–15. The earliest spring record was for February 10. The fall migration through Nebraska extends from mid-September to early November, with a peak during the period October 11–25. Another analysis by Austin and Richert (2001) determined a median spring arrival data of April 12 for Kansas and Nebraska, and April 19 for South Dakota, Median fall arrival dates were October 23 for South Dakota, and October 27 for Kansas and Nebraska.

Habitats: While in Nebraska, the central Platte Valley is the primary habitat, and a wide and slow-flowing river, with its numerous sand bars and islands, and adjacent wet meadows, grain fields and marshlands, is evidently
an important combination of habitat characteristics. This species migrates
later in spring than does the sandhill crane, and thus does not normally associate with it. It uses marshy areas, playa "lagoons," and similar wet areas
for foraging to a larger degree than does the sandhill crane.

Comments: Whooping cranes have stopped in Nebraska's Platte Valley more often than anywhere else along their entire migration route be-

tween wintering and breeding areas. Unlike sandhill cranes, flock sizes are quite small, often comprised of single extended families. These sometimes contain individuals representing as many as four successive generations, of which the oldest may be more than 30 years of age (Gill & Johnsgard, 2010). In recent years a few whooping cranes have often arrived early with sandhill cranes, presumably these are often young birds that have been separated from their families and joined wintering sandhill crane flocks. Recent populations of the historic Wood Buffalo Park–Aransas N.W.R. flock have approached 300 birds, and a few hundred other birds exist in captivity or experimentally established flocks..

Suggested Reading: B.O.N.A. 153; Gill & Johnsgard, 2010, Johnsgard, 2011.

ORDER CHARADRIIFORMES - SHOREBIRDS

Family Charadriidae–Plovers

Black-bellied Plover, *Pluvialis squatarola*

An uncommon spring and fall migrant in eastern parts of the region, becoming rarer westward. Less common in fall than during spring, but more common in both seasons than the golden-plover.

Migration: Sixty-six total spring sightings in Nebraska range from April 4 to June 9, with a median of May 16. Half of the records fall within the period May 12–23. Thirteen initial fall sightings are from July 27 to October 2, with a median of August 20. Thirteen final fall sightings are from August 27 to November 12, with a median of October 6. The birds are most numerous at Cheyenne Bottoms Wildlife Area throughout May, and again from late August until early October (Zimmerman, 1991).

Habitats: Mudflats, shallow ponds and plowed fields are used by migrating birds.

Comments: In contrast to most shorebirds, ploughed fields are a favorite habitat for migrant black-bellied plovers. In these locations the black and white plumage patterns seem appropriate. This species and the golden-plovers are almost unique among American shorebirds in having darker underparts than backs, which is contrary to the principle of countershading for maximum concealment. Jorgensen (2004) observed peak numbers occur-

ring in the eastern Rainwater Basin during the second week of May, with total birds seen annually during the five-year study period ranging from 74–
528. The North American population has been estimated at 200,000 birds
(Morrison *et al.*, 2001).

Suggested Reading: Johnsgard, 1981; Hayman *et al.*, 1986; B.O.N.A.
186; Byrkjedal and Thompson, 1998.

American Golden-Plover, *Pluvialis dominica*

An uncommon to occasional migrant in eastern part of the region, rarer
westwardly. More common in spring than fall, but present both seasons.

Migration: Forty-nine total spring sightings in Nebraska are from April 6
to May 29, with a median of May 7. Half of the records fall within the period
April 25 to May 14. Ten initial fall sightings are from September 2 to October 9, with a median of September 28. Ten final fall sightings are from September 8 to November 20, with a median of October 12. Sharpe, Silcock and
Jorgensen (2001) noted that spring reports extend from March 22 to June
5, and fall migration reports are from July 11 to November 20.

Habitats: Migrants favor grass stubble, short pasturelands, and newly
plowed fields.

Comments: In common with the previous species, the upperpart coloration of adults in breeding plumage is spangled with dark and light markings. When crouched on a nest these patterns merge almost perfectly with
the wet tundra (golden-plover) or dry tundra (black-bellied plover) habitats
that are their respective preferred nest sites. Adults leave their arctic nesting grounds in August, and the young in September, with both age-groups
moving southeast to Canada's maritime provinces and the coast of New
England. They then make an apparently nonstop overseas flight to South
America, wintering in grasslands from Bolivia to Argentina. In spring they
head north through the interior of the U.S., reaching maximum numbers at
Cheyenne Bottoms in mid-May (Zimmerman, 1991). A few birds also migrate south through interior North America in the fall. Jorgensen (2004)
observed peak numbers occurring in the eastern Rainwater Basin during
the third week of May, with total birds seen annually during the five-year
study period ranging from 228–1,065. Three years of fall counts varied from
34–160 total birds. The North American population has been estimated at
150,000 birds (Morrison *et al.*, 2001).

Suggested Reading: Johnsgard, 1981; Hayman *et al.*, 1986; B.O.N.A.
201; Byrkjedal and Thompson, 1998.

Semipalmated Plover, *Charadrius semipalmatus*

An uncommon to occasional spring and fall migrant throughout the region, but probably more common eastwardly. It also migrates throughout the entire Great Plains region.

Migration: Eighty-two initial spring sightings in Nebraska are from March 24 to June 6, with a median of May 12. Sixteen initial fall sightings are from July 25 to September 24, with a median of August 11. Sixteen final fall sightings are from July 30 to October 14, with a median of September 18. Sharpe, Silcock and Jorgensen (2001) noted that spring reports extend from March 24 to June 10, and fall migration reports are from July 3 to October 20.

Habitats: Migrants favor mudflats, shallow ponds, and the muddy banks of slowly flowing rivers.

Comments: Unlike the snowy and piping plovers, this species is a high-arctic nester, nesting on pebbly tundra sites. Its back is the color of fairly dark, wet sand, which helps to distinguish the species from the two just-mentioned species. Jorgensen (2004) observed peak numbers occurring in the eastern Rainwater Basin during the fourth week of April, with total birds seen annually during the five-year study period ranging from 69–507. The North American population has been estimated at 150,000 birds (Morrison *et al.*, 2001).

Suggested Reading: Johnsgard, 1981; Hayman *et al.*, 1986; B.O.N.A. 444.

Family Scolopacidae–Sandpipers and Phalaropes

Greater Yellowlegs, *Tringa melanoleuca*

A common spring and fall migrant throughout the region.

Migration: The range of 115 initial spring sightings in Nebraska is from March 13 to June 10, with a median of April 13. Half of the records fall within the period April 2–14. The range of 55 final spring sightings in Nebraska is from April 11 to May 30, with a median of May 5. Thirty-eight initial fall sightings are from July 20 to October 16, with a median of August 18. Half of the records fall within the period August 4 to September 3. Thirty-eight final fall sightings are from August 14 to November 16, with a median of October 7. Sharpe, Silcock and Jorgensen (2001) noted that spring reports extend from March 3 to May 31, with a spring peak in April, and fall migration reports are from June 10 to November 18.

Habitats: Ponds, marshes, creeks, mud flats and flooded meadows are used by migrants.

Comments: Somewhat larger than the lesser yellowlegs, this species has a considerably longer and more robust bill, and when taking flight it usually utters three or four short notes, rather than the two-noted call typical of lesser yellowlegs. The greater yellowlegs is generally less common in the region than is the lesser. Jorgensen (2004) observed peak numbers occurring in the eastern Rainwater Basin during the first week of April, with total birds seen annually during the five-year study period ranging from 94–451. Three years of fall counts varied from 66–310 total birds annually. Morrison *et al.* (2001) estimated the North American population at 100,000 birds.

Suggested Reading: Johnsgard, 1981; Hayman *et al.*, 1986; B.O.N.A. 355.

Solitary Sandpiper, *Tringa solitaria*

A common to occasional spring and fall migrant throughout the region, being most abundant eastwardly, and probably least common in the treeless areas. This species' status and distribution in Nebraska has been analyzed in some detail (*Nebraska Bird Review* 10:15-22).

Migration: Eighty-eight initial spring sightings in Nebraska are from March 17 to June 7, with a median of May 4. Half of the records fall within the period April 28 to May 11. Twenty-nine final spring sightings in Nebraska are from May 6 to June 10, with a median of May 13. Thirty-six initial fall sightings are from July 20 to September 9, with a median of August 9. Thirty-five final fall sightings are from August 5 to November 26, with a median of September 1. Sharpe, Silcock and Jorgensen (2001) noted that spring reports extend from March 27 to May 9, and fall migration reports are from June 24 to November 2.

Habitats: Wooded ponds, streams and flooded meadows are used by migrants.

Comments: This is indeed a "solitary sandpiper," since it typically forages alone rather than in groups of its own species or even near other shorebirds. Often it can be found along small creeks in the Sandhills that are lined with bushes or trees. Jorgensen (2004) observed peak numbers occurring in the eastern Rainwater Basin during the first week of May, with total birds seen annually during the five-year study period ranging from 3–14. Three years of fall counts varied from 21–124 total birds annually. The North American population has been estimated at 25,000 birds (Morrison *et al.*, 2001).

Suggested Reading: Johnsgard, 1981; Hayman *et al.*, 1986; B.O.N.A. 156.

Lesser Yellowlegs, *Tringa flavipes*

A common spring and fall migrant throughout the region, usually some-what more abundant than the greater yellowlegs.

Migration: The range of 124 initial spring sightings in Nebraska is from March 13 to May 29, with a median of April 14. Half of the records fall within the period April 10 to June 1, with a median of May 13. Thirty-five initial fall sightings are from July 20 to September 22, with a median of August 15. Half of the records fall within the period August 8–September 5. Forty-two final fall sightings are from August 20 to November 23, with a median of October 5. Sharpe, Silcock and Jorgensen (2001) noted that spring reports extend from March 10 to June 6, with a peak the latter half of April , and fall migration reports are from June 19 to November 25. As is the case with most arctic-breeding shorebirds, "fall" migration often begins before the start of calendar summer.

Habitats: Ponds, marshes, creeks, mud flats and flooded meadows are used by migrants. There is no apparent habitat separation of migrating greater and lesser yellowlegs.

Comments: Both species of yellowlegs are tundra nesters, and it is always a shock for persons used to seeing these birds only on wintering or migration sites to find them perching in bushes and low trees in the arctic, where they scan for possible predators or other danger, and utter territorial calls. In eastern Nebraska records suggest that the lesser yellowlegs is nearly ten times more common than the greater. Jorgensen (2004) observed peak numbers occurring in the eastern Rainwater Basin during the fourth week of April, with total birds seen annually during the five-year study period ranging from 473–2,758. Three years of fall counts varied from 211–1,401 total birds annually. The North American population has been estimated at 500,000 birds (Morrison *et al.,* 2001), or five times than that of the greater.

Suggested Reading: Johnsgard, 1981; Hayman *et al.,* 1986; B.O.N.A. 427.

Whimbrel, *Numenius phaeopus*

An extremely rare spring migrant. There have been at least eleven Ne-braska sightings since 1949, with the largest number from Lincoln County, but with other sightings in York, Adams, Webster and Lancaster Counties. There are few fall Nebraska records. In South Dakota it is extremely rare in spring, and there is a single fall record (Tallman, Swanson and Palmer, 2002). In Kansas, it is rare in spring and fall (Thompson *et al.,* 2011).

Migration: Eleven spring Nebraska records are from April 12 to May 27, with a median of May 10. Sharpe, Silcock and Jorgensen (2001) noted that about 35 spring reports extend from April 10 to May 30, and five fall migration reports are from July 19 to October 17.

Habitats: Migrating birds favor flooded grasslands, sandbars, and the shorelines of large impoundments.

Comments: There should be more regional records for this species than the summary above would indicate. The birds nest in Canadian tundra directly north of our region, and some winter on the Gulf Coast, so a regular movement through the three-state corridor seems likely. Jorgensen (2004) observed peak numbers occurring in the eastern Rainwater Basin during the second week of May, with total birds seen annually during the five-year study period ranging from none to 11.

Suggested Reading: Johnsgard, 1981; Hayman *et al.*, 1986; B.O.N.A. 219.

Hudsonian Godwit, *Limosa haemastica*

An uncommon spring migrant in eastern parts of the region, becoming rare or absent in the west. There are no fall Nebraska records for South Dakota, and very few for Nebraska or Kansas.

Migration: Sixty-nine initial spring Nebraska records are from April 12 to May 27, with a median of May 2. Half of the records fall within the period April 22 to May 12. Ten final spring sightings in Nebraska are from May 6–25, with a median of May 15. Sharpe, Silcock and Jorgensen (2001) noted that spring reports extend from April 13 to June 6, with a peak in middle to late April, and three fall migration reports are from August 30 to September 22.

Habitats: Associated with marshy ponds, wet grasslands, and flooded fields while on migration.

Comments: From its arctic nesting grounds along Alaska's Beaufort Sea coastline, and Canada's Hudson Bay, these birds fly southeast to James Bay in late July and early August. After a period of intensive foraging there, they make a apparently nonstop flight of about 3,000 miles to South America, eventually wintering from northern Argentina south to as far as Tierra de Fuego. During their return northward flight they pass through the Great Plains, becoming most numerous at Cheyenne Bottoms Wildlife Area from late April until late May. Most leave on a quick flight to the tundra by early June (Zimmerman, 1991). Jorgensen (2004) observed peak numbers occurring in the eastern Rainwater Basin during the second week of May, with total birds seen annually during the five-year study period ranging from 59–300.

Suggested Reading: Johnsgard, 1981; Hayman *et al.*, 1986; B.O.N.A. 629.

Ruddy Turnstone, *Arenaria interpres*

An occasional to rare spring migrant in eastern parts of the region. In Nebraska, turnstones are very rarely seen as far west as Cherry and Garden counties. The largest numbers of sightings are for Adams and Lancaster counties, especially the latter. Rarely reported during the fall, with a total of only 12 Nebraska observations (*Nebraska Bird Review* 73; 140).

Migration: Twenty-three total spring Nebraska records are from April 19 to May 27, with a mean of May 18; Half of the records fall within the period May 14–25, with a median of May 18. Half the records fall between the period May 14–25. Sharpe, Silcock and Jorgensen (2001) noted that spring reports extend from April 29 to June 4, with a likely peak in the third week of May, and ten fall migration reports are from July 19 to October 28.

Habitats: Shallow ponds, sandy or rocky shorelines, and plowed fields are used by migrating birds.

Comments: Turnstones get their vernacular name from their tendency to flick rocks and pebbles over with their sharp bills, to find small invertebrates that may be hiding below. Few occur regionally, but they are more likely to occur on pebbly beaches than muddy shorelines. Jorgensen (2004) observed peak numbers occurring in the eastern Rainwater Basin during the third week of May, with total birds seen annually during the five-year study period ranging from none to 40. The North American population has been estimated at 235,000 birds (Morrison *et al.,* 2001).

Suggested Reading: Johnsgard, 1981; Hayman *et al.,* 1986; B.O.N.A. 537.

Red Knot, *Calidris canutus*

A rare spring and fall migrant, mainly in eastern parts of the region. In Nebraska, many of the records are for Lancaster and Douglas–Sarpy counties, but there are scattered sightings elsewhere in Nebraska, west to Dundy County.

Migration: A total of six spring Nebraska records range from May 7–19, with a mean of May 14. Seven fall Nebraska records are from August 27 to November 1, with a mean of September 12. Sharpe, Silcock and Jorgensen (2001) noted that 11 spring reports extend from April 30 to May 26, and five fall reports are from July 19 to October 17. Twelve documented fall migration reports include seven for August and four for September, and collectively range from August 4 to October 30.

Habitats: Mud flats and sandbars are used by migrating birds.

Comments: This is a rather robust little shorebird; its distinctive reddish underparts in spring is the basis for its common name "robin snipe" along the Atlantic coast, where it is an abundant migrant. It is a high-arctic nester, and its short bill would seem better adapted for surface foraging than for probing on mud.

Suggested Reading: Johnsgard, 1981; Hayman *et al.*, 1986; B.O.N.A. 563.

Sanderling, *Calidris alba*

A rare to occasional spring and fall migrant throughout the region, mostly in eastern and central parts, and becoming rare in the far west. It is a regular migrant throughout the eastern half of the Great Plains region.

Migration: Fifty-six initial spring sightings in Nebraska are from March 26 to June 2, with a median of May 6. Half of the records fall within the period April 25 to May 15. Thirteen final spring sightings in Nebraska are from April 26 to June 10, with a median of May 13. Seventeen initial fall sightings are from July 27 to October 2, with a median of August 20. Twelve final fall sightings are from August 12 to October 19, with a median of October 4. Sharpe, Silcock and Jorgensen (2001) noted that spring reports extend from March 26 to June 4, and five fall migration reports are from July 1 to November 10.

Habitats: Migrants are associated with sandy shorelines, sandy river bars, salt-encrusted flats, and less frequently muddy shorelines.

Comments: When sanderlings pass through the central Great Plains they mostly are in the pale winter plumage, during which time their contrasting darker anterior wing-coverts provide conspicuous field-marks. On their arctic nesting grounds they have a strong rufous cast on the upperparts. Sanderlings are another of the long-distance shorebird migrants, breeding above the Arctic Circle and wintering south to Tierra del Fuego. Jorgensen (2004) observed peak numbers occurring in the eastern Rainwater Basin during the second week of May, with total birds seen annually during the five-year study period ranging from 3–187. The North American population has been estimated at 300,000 birds (Morrison *et al.*, 2001).

Suggested Reading: Johnsgard, 1981; Hayman *et al.*, 1986; B.O.N.A. 653.

Semipalmated Sandpiper, *Calidris pusilla*

A common spring and fall migrant throughout the Great Plains region, locally abundant or uncommon.

Migration: Eighty-nine initial spring sightings in Nebraska are from March 21 to June 10, with a median of April 28. Half of the records fall within the period April 20 to May 10. Thirty-nine final spring sightings in

Nebraska are from April 28 to June 1, with a median of May 15. Twenty-three initial fall sightings are from July 20 to September 8, with a median of August 5. Twenty-three final fall sightings are from July 28 to October 16, with a median of September 18. Sharpe, Silcock and Jorgensen (2001) noted that spring reports extend from March 21 to June 10, and fall reports are from July 4 to October 10. Spring migration peaks in late April or early May, and most fall migration reports are from the second half of July through the first week of September.

Habitats: Migrating birds use mud flats, shallow ponds, exposed sand bars, and open shorelines as well, but rarely move onto dry fields with Baird's sandpipers or wet grasslands with least sandpipers.

Comments: Like the Baird's sandpiper, this species has a semiannual migration pattern. From their breeding grounds in the western arctic of North America and Asia, the adults arrive at Cheyenne Bottoms Wildlife Area in late July, the females ahead of the males, and the juveniles arriving in August. By late July the birds are present in large numbers, which persist until mid-September. They then continue south to the Gulf Coast, from which they make a relatively easy nonstop flight to the coast of Venezuela. Most then continue on to the coast of Surinam in northeastern South America, where about 1.4 million spend the winter. The return flight is similar to the fall pattern, with maximum spring numbers at Cheyenne Bottoms Wildlife Area occurring from the last week in April through the first week in May (Zimmerman, 1991). Jorgensen (2004) found this species to be the third most abundant spring migrant shorebird in Nebraska's eastern Rainwater Basin. He observed peak numbers occurring in the eastern Rainwater Basin during the second week of May, with total birds seen annually during the five-year study period ranging from 933–5,254. Three years of fall counts varied from 59–476 total birds. The North American population has been estimated at 3.5 million birds (Morrison *et al.*, 2001).

Suggested Reading: Johnsgard, 1981; Hayman *et al.*, 1986; B.O.N.A. 6.

Western Sandpiper, *Calidris mauri*

An uncommon spring and rare fall migrant in eastern parts of the region, becoming more common westward. It is a migrant throughout the Great Plains region, but is more common in spring.

Migration: Forty-one initial spring sightings in Nebraska are from April 7 to June 10, with a median of May 8. Half of the records fall within the period April 28 to May 15. Ten final spring sightings in Nebraska are from May 3–23, with a median of May 13. Fourteen initial fall Nebraska records are from July 20 to September 19, with a median of August 12. Eleven final fall sightings are from August 26 to October 2, with a median of September 1. Sharpe, Silcock and Jorgensen (2001) noted that about spring reports extend from April 12 to May 19, and five fall reports are from July 4 to October 29. The peak of spring migration occurs during late April. The fall numbers are larger, and probably peak in late August, when many juveniles are present.

Habitats: Mud flats, shallow ponds and open shorelines are used by migrants, which avoid dry areas and prefer to forage while wading in shallow water, usually forage a slightly greater depth than do semipalmated sandpipers.

Comments: Western sandpipers are much like semipalmated sandpipers in appearance, but are more rufous above (especially in spring) and have a longer bill that droops slightly at the tip. However, their migration routes are very different. Western sandpipers are relatively rare in spring on the Great Plains, but are abundant during the fall. They are most abundant at Cheyenne Bottoms Wildlife Area from mid-July until the third week of September. They arrive there by migrating southeast from Alaska's North Slope, or flying more directly east from the Gulf of Alaska and coastal British Columbia. From Cheyenne Bottoms Wildlife Area the birds continue southeast to Florida, cross the Gulf of Mexico, and winter along the northeastern coast of South America. Probably the great majority take a more directly migration south along the Pacific coast to the coast of northern Peru (Zimmerman, 1991). Virtually all of the birds migrate north in spring along the Pacific coast, producing enormous flocks of migrants at a few staging areas, such as along Alaska's Copper River delta. The North American population has been estimated at 3.5 million birds (Morrison *et al.,* 2001).

Suggested Reading: Johnsgard, 1981; Hayman *et al.,* 1986; B.O.N.A. 90.

Least Sandpiper, *Calidris minutilla*

A common spring and fall migrant throughout the region, but becoming less common westward.

Migration: The range of 102 initial spring sightings in Nebraska is from March 8 to May 29, with a median of May 2. Half of the records fall with the period April 20 to May 10. Forty-one final spring sightings in Nebraska are from April 27 to June 2, with a median of May 12. Twenty-three final fall sightings are from July 20 to September 9, with a median of August 2. Twenty-three final fall sightings are from July 28 to November 11, with a median of September 18. Sharpe, Silcock and Jorgensen (2001) noted that spring reports extend from March 7 to June 10, and fall migration reports are from June 23 to November 23. The spring migration peak is from the last week in April through the first two weeks of May. August is the peak of fall migration.

Habitats: Mud flats, shallow ponds, marsh edges and flooded meadows are used by migrants, which frequently gather in small groups foraging in shallow puddles or wet grasslands usually well away from the larger "peeps."

Comments: This is one of the commonest "peeps" in the region, and is notable for its olive-yellow legs and small size. Jorgensen (2004) observed peak numbers occurring in the eastern Rainwater Basin during the first week of May, with total birds seen annually during the five-year study period ranging from 362–1,545. Three years of fall counts varied from 270–802 total birds. Morrison *et al.* (2001) estimated the North American population at 600,000 birds.

Suggested Reading: Johnsgard, 1987; Hayman *et al.*, 1986; B.O.N.A. 115.

White-rumped Sandpiper, *Calidris fuscicollis*

An abundant spring migrant throughout the region, but few if any fall valid records exist. It is probably somewhat more common in eastern than in western areas.

Migration: The range of 100 initial spring sightings in Nebraska is from March 28 to June 1, with a median of April 29. Half of the records fall within the period May 1-16. Seventeen final spring sightings in Nebraska are from May 8 –25. Sharpe, Silcock and Jorgensen (2001) noted that spring reports extend from April 19 to June 21, with large numbers present from middle to late May. The fall migration, like that of the Hudsonian godwit, occurs along the Atlantic Coast, so no white-rumped sandpipers are likely to be seen in the central Great Plains during fall.

Habitats: Migrants feed in shallow ponds, flooded pastures, flat shorelines, and muddy creeks; often with Baird's sandpipers, but are less likely to forage in dry areas than that species, and are more prone to be mixed with semipalmated sandpipers.

Comments: This species differs from the closely related Baird's sandpiper in being polygynous rather than monogamous, and in having an elliptical or loop-like migration pattern. After the females begin incubation, the males depart the breeding areas, and forage in preparation for the long fall migration. They are joined by females after the young hatch, and all the birds head for the Atlantic coast of Canada. From there they make a nonstop flight to Surinam, and then continue south along South America's Atlantic coast to wintering sites from Paraguay and northern Argentina south to Tierra del Fuego. The spring flight brings them through the Great Plains, where they become abundant at Cheyenne Bottoms Wildlife Area from mid-May to early June. At that time they make a single nonstop flight back to their tundra breeding grounds (Zimmerman, 1991). Jorgensen (2004) found it to be the most abundant spring migrant shorebird in the eastern Rainwater Basin. He observed peak numbers occurring in the eastern Rainwater Basin during the third week of May, with total birds seen annually during the five-year study period ranging from 3,304–8,983. The North American population has been estimated at 400,000 birds (Morrison *et al.*, 2001).

Suggested Reading: Johnsgard, 1981; Hayman *et al.*, 1986; B.O.N.A. 29.

Baird's Sandpiper, *Calidris bairdii*

A common spring and fall migrant throughout the region, and probably the most abundant of the regional "peeps", especially in western areas.

Migration: The range of 125 initial spring sightings in Nebraska is from March 12 to May 24, with a median of April 21. Half of the records fall within the period April 6 to May 4. Fifty-four final spring sightings in Nebraska are from April 7 to May 29, with a median of May 13. Thirty-two initial fall sightings are from July 20 to October 1, with a median of August 12. Twenty-seven final fall sightings are from August 3 to December 5, with a median of October 6. Sharpe, Silcock and Jorgensen (2001) noted that spring reports extend from March 8 to June 7, and fall reports are from July 9 to November 8. Spring arrival is relatively early, with birds common through April. Fall numbers are fewer, and these birds are more likely to be found in western areas, such as at Lake McConaughy.

Habitats: Migrants use mud flats, shallow ponds, sand bars, and dried areas such as overgrazed pastures, salt plains and similar open habitats while on migration.

Comments: Like most of the migrant sandpipers, the Baird's is semiannual in its movements through the Great Plains to nesting areas above the Arctic Circle. The adult birds leave their nesting grounds early, by late June and early July, the females in advance of the males. Their flight south takes them through eastern Colorado and western parts of Nebraska and Kansas. The juveniles migrate later, in late August, and migrate over a broader pathway, with many moving as far east as the Atlantic coast. Post-breeding adults begin to arrive at Cheyenne Bottoms Wildlife Area in mid-July, with numbers peaking during the first half of August. After staging there, the birds make a nonstop, roughly 4,000-mile, flight to the Andes of northern South America. Young birds make a similarly long flight, from staging areas in the American Southwest to western South America. From the northern Andes the birds continue south to wintering areas as far as Tierra del Fuego. The return flight takes a similar route, with the birds staging at Cheyenne Bottoms Wildlife Area from late March to the middle of May, then rushing to their arctic breeding grounds (Zimmerman, 1990). Jorgensen (2004) found this species to be the sixth most abundant spring migrant shorebird in the eastern Rainwater Basin. He observed peak numbers occurring in the eastern Rainwater Basin during the last two weeks of April, with total birds seen annually during the five-year study period ranging from 881–2,001. The North American population has been estimated at 300,000 birds (Morrison et al., 2001).

Suggested Reading: Johnsgard, 1981; Hayman *et al.*, 1986; Zimmerman, 1990; B.O.N.A. 661.

Pectoral Sandpiper, *Calidris melanotos*

A common to abundant spring and fall migrant almost throughout the region, but becoming less common to rare westward.

Migration: The range of 102 initial spring sightings in Nebraska is from March 4 to June 6, with a median of April 28 and half of the records falling within the period April 15 to May 8. Thirty-nine final spring sightings in Nebraska are from April 5 to May 25, with a median of May 13. Twenty-eight fall sightings are from August 3 to November 20, with a median of October 4. Sharpe, Silcock and Jorgensen (2001) noted that spring reports extend from March 4 to June 8, and fall reports are from July 5 to November 9. Spring peak migration is from the last week of April through the first two weeks of May, and during fall occurs in August and September.

Habitats: Migrating birds use a variety of habitats, including muddy shorelines, creeks, flooded grasslands and shallow marshy areas where the emergent vegetation is not too thick.

Comments: Larger than the typical "peeps," the pectoral sandpiper often feeds among them, when its greater size and sharply cut-off breast pattern is usually quite apparent. Jorgensen (2004) found this species to be the most abundant fall migrant shorebird in the eastern Rainwater Basin. Zimmerman (1991) notes that it s most common at Cheyenne Bottoms Wildlife Area through May, and becomes abundant again in late July and early August. Even those birds using Cheyenne Bottoms in spring and breeding in central Siberia migrate east in late summer across the Bering Strait to join those that bred in North America. Jorgensen (2004) observed peak numbers occurring in the eastern Rainwater Basin during the first week of

May, with total birds seen annually during the five-year study period rang-ing from 537–1,492. Three years of fall counts varied from 568–1,082 total birds. The North American population has been estimated at 400,000 birds (Morrison *et al.*, 2001).

Suggested Reading: Johnsgard, 1981; Hayman *et al.*, 1986; B.O.N.A. 348.

Dunlin, *Calidris alpina*

An occasional spring migrant in eastern parts of the region, rarer in west-ern areas. Rare during fall migration in all parts of the region.

Migration: Forty-eight spring sightings in Nebraska range from April 6 to June 2, with a median of May 13. Half of the records fall within the pe-riod May 9–21. Eleven fall Nebraska records are from August 15 to Novem-ber 20, with a median of September 11. It has been observed as far west as Cherry and Garden counties, but is extremely rare in western areas. Sharpe, Silcock and Jorgensen (2001) noted that spring reports extend from April 3 to June 9, and fall reports are from September 4 to November 26. The spring peak movement is during May, and in fall occurs during the latter half of October.

Habitats: Migrants use mud flats, shallow ponds, and open stretches of muddy shorelines, often mingling with other small sandpipers.

Comments: Like the black-bellied plovers and golden-plovers, dunlins in breeding plumage have conspicuous black bellies, a most unexpected fea-ture for a tundra nester. Perhaps this feature is displayed during territorial flights, or otherwise has some special signal value. Jorgensen (2004) ob-served peak numbers occurring in the eastern Rainwater Basin during the third week of May, with total birds seen annually during the five-year study period ranging from 30–159. The North American population has been esti-mated at 1.5 million birds (Morrison *et al.*, 2001).

Suggested Reading: Johnsgard, 1981; Hayman *et al.*, 1986; B.O.N.A. 203.

Stilt Sandpiper, *Calidris himantopus*

A common or uncommon spring and fall migrant almost throughout the region, but becoming less abundant westward.

Migration: Ninety-nine initial spring sightings in Nebraska are from April 3 to May 29, with a median of May 11. Half of the records fall within the period May 9–19. Sixteen final spring sightings in Nebraska are from May 7–30, with a median of May 17. Eleven initial fall sightings are from July 21 to September 19, with a median of August 11. Nine final fall sightings are from September 3 to October 21, with a median of September 20. Sharpe, Silcock and Jorgensen (2001) noted that spring reports extend from April 14 to June 1, and fall reports are from July 5 to October 27. The peak of spring migration is during May, with about two-thirds of the reports from May 8–23. The peak of fall migration is during August and September. This is a fairly frequently encountered shorebird in the region.

Habitats: Muddy flats, shallow mud-bottom ponds and flooded fields are used by migrants; the birds feed in belly-deep water and are more likely to be in sheltered areas than on exposed shorelines than many other shorebirds

Comments: Like the pectoral sandpiper, stilt sandpipers are most common throughout May at Cheyenne Bottoms Wildlife Area, and again are abundant there in late July and early August, migrating to and from tundra breeding grounds. Jorgensen (2004) found this species to be the fifth most abundant spring migrant shorebird in Nebraska's eastern Rainwater Basin. He observed peak numbers occurring in the eastern Rainwater Basin during the second week of May, with total birds seen annually during the five-year study period ranging from 824–4,388. Three years of fall counts varied from 84–599 total birds. The North American population has been estimated at 200,000 birds (Morrison *et al.,* 2001).

Suggested Reading: Johnsgard, 1981; Hayman *et al.,* 1986; B.O.N.A. 341.

Buff-breasted Sandpiper, *Tringites subruficollis*

A generally rare spring and fall migrant in the region, but uncommon in eastern Nebraska, which appears to be one of the few major stopover areas during the species' transequatorial migration route.

Migration: Twelve total spring sightings in Nebraska are from May 1–20, with a median of May 10. Eleven fall sightings are from August 17 to September 26, with a median of September 7. Sharpe, Silcock and Jorgensen (2001) noted that about spring reports extend from April 23 to May 25, and five fall reports are from July 17 to September 26. Most birds don't arrive until about May 10, with high counts of often more than 100 birds. Fall migration begins by early August and continues into mid-September, with the migration pathway wider in fall than spring. The species is apparently very rare in western Nebraska, but has been reported from Scotts Bluff and Sheridan counties.

Habitats: Migrants are usually found on recently plowed fields, mowed or burned grasslands, meadows, heavily grazed pastures and other rather dry habitats. This species is highly local, but has been reported several times in

York and Seward counties. Jorgensen (2004) found this species to be regular in the eastern Rainwater Basin, especially in Seward and Fillmore counties. He observed that the birds forage in agricultural fields, hayfields and wetlands. He saw maximum numbers during the second and third weeks of May, and determined that Nebraska is probably the species' most important spring staging area, supporting a large percentage of the world's population in May. Jorgensen observed numbers in the eastern Rainwater Basin to drop of quickly after the fourth week of May, with total birds seen annually during the five-year study period ranging from 136–459. During fall the birds move south thorough the Great Plains, Mexico, Central America and South America along the east side of the Andes to wintering grounds in northern Argentina and southern Paraguay. The return trip follows roughly the same route.

Comments: This species' total population has been estimated to perhaps be as low as about 15,000-20,000 birds, probably all of which passes through central Nebraska. The rather limited area of the eastern Rainwater Basin used by the birds in spring makes it of special conservation significance for this little-studied and possibly threatened species.

Suggested Reading: Johnsgard, 1981; Hayman *et al.,* 1986; B.O.N.A. 91; Jorgensen, 2004.

Short-billed Dowitcher, *Limnodromus griseus*

A rare migrant in eastern parts of the region, becoming rarer to the west. The species is probably regular only in the northeastern portions of the Great Plains region.

Migration: Seven total spring sightings in Nebraska attributed to this species are from April 20 to May 18, with a median of May 14. Fourteen fall sightings are from early August to September 10, with most records between August 19 and September 10. Sharpe, Silcock and Jorgensen (2001) noted that spring reports extend from April 25 to May 23, and fall reports are from August 5 to September 11. The peak of spring migration is during mid-May, and most of the fewer fall records are of juveniles. Museum studies of dowitchers in Nebraska (*Nebraska Bird Review* 8:63-74; 64:74-78) suggest that the short-billed dowitcher is fairly rare in the state. Most recent sight records are from eastern counties (*e.g.,* Jorgensen, 2004).

Habitats: Migrants use muddy flats and mud-bottom ponds that are probably identical to those of the long-billed dowitcher. They are often seen with black-bellied plovers.

Comments: Rather little is known of this species' regional occurrence. In spring, it is less rufous below than is the long-billed dowitcher, and its usual call is a three-noted whistle, but bill-length differences in the two species are not apparent in the field. Jorgensen (2004) observed peak numbers occurring in the eastern Rainwater Basin during the second and third weeks of May, with total birds seen annually during the five-year study period ranging from 12–106. The North American population has been estimated at 320,000 birds (Morrison *et al.*, 2001).

Suggested Reading: Johnsgard, 1981; Hayman *et al.*, 1986; B.O.N.A. 564.

Long-billed Dowitcher, *Limnodromus scolopaceus*

A common spring and fall migrant throughout the region.

Migration: Thirty-five initial spring sightings in Nebraska range from April 12 to May 23, with a median of May 1. Half of the records fall within the period April 20 to May 11. Thirteen final spring sightings in Nebraska are from May 4 to June 1, with a median of May 11. Eleven initial fall sightings are from July 20 to October 7, with a median of August 8. Thirteen final fall sightings are from August 1 to December 3, with a median of October 14. Sharpe, Silcock and Jorgensen (2001) noted that spring reports extend from March 18 to June 13, and fall reports are from July 19 to November 19. The peak of spring migration is from the last week of April through the first half of May; during fall they might be seen from August well into October. Unless carefully identified, dowitchers in the region should be tentatively assigned to this species. Many birds are nearly impossible to identify during fall.

Habitats: Associated with muddy flats and mud-bottom ponds in Nebraska; foraging is done by probing in shallow water of ponds or flooded grasslands.

Comments: Most dowitchers seen in the region are of this species, which is distinguished by its rich rufous underparts and usually uttering a single *keek* note upon takeoff. Jorgensen (2004) found this species to be the fourth most abundant spring migrant shorebird in the eastern Rainwater Basin, and the second most abundant during the fall. Zimmerman (1991) noted that long-billed dowitchers are most common at Cheyenne Bottoms Wildlife Area throughout April and until mid-May, and from late July to the first week in November. Jorgensen (2004) observed peak numbers occurring in the eastern Rainwater Basin during the first week of May, with total birds seen annually during the five-year study period ranging from 1,788–3,100. From 215–1,027 total birds were seen annually during three years of fall counts. The North American population has been estimated at 500,000 birds (Morrison *et al.*, 2001).

Suggested Reading: Johnsgard, 1981; Hayman *et al.*, 1986; B.O.N.A. 493; Jorgensen, 2004.

Red-necked Phalarope, *Phalaropus lobatus*

An uncommon to rare spring migrant in western parts of the region. Less common during fall in all areas.

Migration: Forty-two initial spring sightings in Nebraska range from April 19 to May 27, with a median of May 14. Half of the records fall within the period May 8–19. Seven final spring sightings in Nebraska are from May 9–25, with a mean of May 19. Ten initial fall sightings are from July 20 to September 21, with a median of August 10. Eleven final fall sightings are from August 20 to October 14, with a median of September 27. Sharpe, Silcock and Jorgensen (2001) noted that spring reports extend from April 19 to June 1, and fall reports from July 25 to October 14. The peak of spring migration occurs during the third week of May, when most birds concentrate in the saline wetlands of the northern Nebraska panhandle.

Habitats: Migrants use the same habitats as do Wilson's phalaropes, namely open wetlands of marshes and shallow lakes, where the invertebrate life is abundant and can be captured by surface foraging.

Comments: Once called the northern phalarope, this is an arctic-nesting bird that is only seen in Nebraska while on migration: In non-breeding plumage it often closely resembles the Wilson's phalarope, but is somewhat darker dorsally. Jorgensen (2004) observed peak numbers occurring in the eastern Rainwater Basin during the fourth week of May, with total birds seen annually during the five-year study period ranging from none to 24. The North American population has been estimated at 2.5 million birds (Morrison *et al.*, 2001).

Suggested Reading: Johnsgard, 1981; Hayman *et al.*, 1986; B.O.N.A. 538.

Family Laridae–Gulls and Terns

Sabine's Gull, *Xema sabini*

A rare to very rare fall migrant in Nebraska and Kansas, and accidental in South Dakota. There were at least nine records for this species in Nebraska through 1996, but in 1997 at least 24–30 birds were seen between September 8 and October 12 *(Nebraska Bird Review* 65:167–8). Sharpe, Silcock and Jorgensen (2001) noted that reports have extended from September 3 to October 15, and have mostly consisted of juveniles. It has also been reported at least seven times in South Dakota (Tallman, Swanson and Palmer, 2002), and from at least 24 Kansas counties (Thompson *et al.*, 2011).

Habitats: Large rivers, lakes and reservoirs in the interior are sometimes used, but most migration occurs coastally.

Comments: Unlike some of the rare gulls of Nebraska, this species can be easily identified by the white triangular patch on its upper wing, bounded by all-black wingtips and a black bill with a distinctive yellow tip. It has a circumpolar distribution, and was named for one of the members of the Ross expedition of 1818 while searching for the North West Passage. There are no national population estimates, but the Alaska breeding population may be in the tens of thousands, and about 100,000 may winter along the Pacific coast (B.O.N.A. 593).

Suggested Reading: Richards, 1990; B.O.N.A. 593; Howell and Dunn, 2007.

Bonaparte's Gull, *Choricocephalis philadelphia*

An uncommon spring and fall migrant in eastern parts of the region, becoming rarer westwardly. Migrants are most regular in eastern and northern areas, mainly from the Missouri River eastward.

Migration: Thirty-six total spring sightings in Nebraska are from April 3 to May 27, with a median of April 23. Half of the records fall within the period April 12 to May 9. Twenty fall sightings are from August 18 to November 21, with a median of October 26. There are at least seven January records (*Nebraska Bird Review* 74:10). Sharpe, Silcock and Jorgensen (2001) noted that spring reports extend from March 8 to June 1, and fall reports from September 1 to December 24. The peak of spring migration is during middle to late April, and the fall migration peak is from early to middle November. A state-level analysis of four decade-long periods of Christmas Bird Counts (1967-68 to 2006–7) extending from North Dakota to the Texas panhandle indicated a late-December population peak in Oklahoma (Johnsgard and Shane, 2009).

Habitats: Migrants are associated with rivers, reservoirs, lakes and marshes, especially large lakes.

Comments: This is an arctic-nesting gull whose white outer wing patches make it easily separable from Franklin's gulls, with which it sometimes associates while on migration. Like the Franklin's and some other gulls that often feed on crustaceans, it shows a pink blush of feathers on the underparts in spring. An estimate of its total population in the 1990's was 85,000–175,000 pairs (B.O.N.A., 634).

Suggested Reading: Richards, 1990; B.O.N.A. 634; Howell and Dunn, 2007.

Little Gull, *Hydrocoelus minutus*

A rare migrant in Nebraska and Kansas, and accidental in South Dakota. Observed at Wehrspan Lake on April 26, 1995 (*Nebraska Bird Review* 64:134), and at Pawnee Lake, Lancaster County, October 3, 1996 (Brogie, 1997). It was also seen at North Platte N.W.R. September 6, 1997, and at Summit Reservoir on October 19, 1997 *(Nebraska Bird Review* 65:167). Through 2005 there had been 16 records, mostly during fall (*Nebraska Bird Review* 73:141). Sharpe, Silcock and Jorgensen (2001) noted that two spring reports are for April 19–26, and eight fall reports from September 6 to November 1. It has also been reported at least five times in South Dakota, in June, September, November and December (Tallman, Swanson and Palmer, 2002), and from at least 14 Kansas counties (Thompson *et al.*, 2011).

Habitats: Migrants are associated with rivers, reservoirs, lakes and other wetlands, especially large lakes and reservoirs.

Comments: This is the smallest of all the world's gulls. Although a Eurasian species, breeding from Scandinavia east to Siberia, there have been a few North American nesting records for little gulls, scattered from Hudson Bay south to the Great Lakes region. Nesting was first observed in the Toronto region of Ontario in 1962, and in the U.S. in 1975. However, this colonizing activity has apparently not expanded since the 1980's.

Suggested Reading: Richards, 1990; B.O.N.A. 428; Howell and Dunn, 2007.

Laughing Gull, *Leucophaeus atricilla*

A rare migrant in Nebraska and Kansas, and accidental in South Dakota.

Migration: Five spring Nebraska records are from April 5 to May 21, with a mean of April 22. The species has also been observed in June and July, and three fall Nebraska records are from December 5–22. By 2010 there were 24 state records (*Nebraska Bird* Review 78:139). Sharpe, Silcock and Jorgensen (2001) noted that about 13 spring reports extend from April 2 to May 26. There are also four summer reports, and three fall reports, from October 28 to December 22. There are also records from at least 23 Kansas counties (Thompson *et al.*, 2011) and from two South Dakota counties (Tallman, Swanson and Palmer, 2002).

Habitats: Normally associated with coastal habitats while wintering and on migration, migrants are likely to be seen near large impoundments in the interior.

Comments: This species breeds from the eastern coast of North America south along the Gulf of Mexico, through eastern Mexico and Central America to northern South America, as well as in the West Indies. It has been increasingly reported in the midwestern states. It is slightly larger than the quite similar Franklin's gull, and has a noticeably heavier and somewhat drooping bill. The North American population was estimated at about 26,000 pairs in the 1990's (B.O.N.A. 225).

Suggested Reading: Richards, 1990; B.O.N.A. 225; Howell and Dunn, 2007.

Mew Gull, *Larus canus*

A very rare spring and fall migrant in Nebraska and Kansas; accidental in South Dakota. As of 1998 there were at least 11 Nebraska records, nearly all from Lake McConaughy, but by 2010 there were 24 state records (*Nebraska Bird Review* 78:139) reflecting the increased activity of birding in the state, especially at Lake McConaughy. It has been reported at least three times in South Dakota, in November and December (Tallman, Swanson and Palmer, 2002). It has also been reported from at least nine Kansas counties (Thompson *et al.,* 2011).

Migration: Sharpe, Silcock and Jorgensen (2001) noted that eight spring reports extend from February 17 to May 11, and there is one fall report for December 1–5. There have since been numerous sightings of this Alaskan and Canadian species at Lake McConaughy, between November 6 and May 1 (*Nebraska Bird Review* 66:42; 72:10). It was observed three years during Christmas Bird Counts at Lake McConaughy between 2000–2001 and 2009–2010.

Habitats: Associated with both coastal and inland habitats while wintering and on migration. Migrants are more likely to be seen near large impoundments in the interior than coastally. There they frequently forage on grasslands, where earthworms are often eaten.

Comments: This gull also occurs widely in Europe, from the Atlantic to the Pacific coasts. It has been called (inappropriately) the "common gull" in Great Britain, and in North America it has also been called the "short-billed gull". The name "mew gull" refers to its unusual voice. It is more prone to nest in inland locations than are other arctic to subarctic gulls, except for the Bonaparte's gull. It similarly often nests in trees, sometimes taking over the nests of corvids. The world population was estimated at about one million pairs in the 1990's (B.O.N.A. 687).

Suggested Reading: Richards, 1990; B.O.N.A. 687; Howell and Dunn, 2007.

Herring Gull, *Larus argentatus*

An uncommon spring and fall migrant throughout the region, locally overwintering. Immatures are sometimes seen during the summer months.

Migration: Forty-seven initial spring Nebraska records range from January 13 to May 13, with a median of March 18. Half of the records fall within the period March 2–April 1. Twenty-seven final spring sightings in Nebraska range from March 5 to May 28, with a median of April 21. Twenty-four initial fall sightings are from July 21 to November 24, with a median of October 26. Eighteen final fall sightings are from August 29 to December 21, with a median of November 28. Sharpe, Silcock and Jorgensen (2001) noted that reports extend September 13 to May 31, together with many summer reports of immatures. Large numbers can be found on reservoirs such as Lake McConaughy just prior to the coldest winter weather. A state-level analysis of four decade-long periods of Christmas Bird Counts (1967-68 to 2006–7) extending from North Dakota to the Texas panhandle indicated a late-December population peak in Oklahoma (Johnsgard and Shane, 2009).

Habitats: Migrating birds are widely distributed over rivers, lakes, reservoirs and other larger wetlands.

Comments: This is the largest of the region's regularly occurring gulls, and also the most common of our very large gulls. National Breeding Bird Surveys between 1966 and 2009 indicate that the species underwent a statistically significant population decline (3.6 percent annually) during that period. The Atlantic Coast population from Maine south was estimated at about 100,000 pairs in the mid-1980's, and several thousand more were then breeding in Canada (B.O.N.A. 124).

Suggested Reading: Richards, 1990; B.O.N.A. 124; Howell and Dunn, 2007.

Thayer's Gull, *Larus thayeri*

A rare but regular fall to spring migrant in Nebraska and Kansas, overwintering locally or occasionally, and an extremely rare or accidental vagrant in South Dakota.

Migration: Reported as early as November 24 and as late as April 15, with most sightings from November through January. As of 1996 there were about 22 reports of this species, all since 1991 (*Nebraska Bird Review* 64:53). At least 20 were seen at Lake McConaughy during the winter of 1997–98 (*Nebraska Bird Review* 66:13). Sharpe, Silcock and Jorgensen (2001) noted that Nebraska reports extend from November 6 to April 23.

It has been reported at least seven times in South Dakota, in March, May, October and December (Tallman, Swanson and Palmer, 2002). It has also been reported from at least 24 Kansas counties (Thompson *et al.*, 2011).

Habitats: Rivers, reservoirs, lakes and coastal shorelines are normally used by migrants.

Comments: This high-arctic Canadian gull is of uncertain taxonomic status. Once considered a subspecies of the herring gull, it was raised to the species level in 1998 by the A.O.U. It has also been considered a subspecies (together with *kumlieni*) of the Iceland gull. Because of identification difficulties, sight records for Thayer's and Iceland gulls are hard to verify. The world population of *thayeri* was estimated at about 6,300 pairs in the 1990's (B.O.N.A. 699).

Suggested Reading: B.O.N.A. 699; Howell and Dunn, 2007.

Iceland Gull, *Larus glaucoides*

A rare overwintering migrant in Nebraska and Kansas; an accidental vagrant in South Dakota. An arctic-breeder nesting in Greenland, Baffin Island and the northern part of Hudson Bay.

Migration: There were 22 Nebraska records through 2009 (*Nebraska Bird Review* 77:27). Sharpe, Silcock and Jorgensen (2001) noted that reports extend through winter, from November 10 to April 20. There are ten mid-winter records, from January 10 to February 18. It has been reported at least three times in South Dakota, during March and December (Tallman, Swanson and Palmer, 2002). It has also been reported from at least five central and eastern Kansas counties, from December to March (Thompson *et al.*, 2011). Most Nebraska records are from western areas, but include Saline and Lancaster counties. It was observed four years during Christmas Bird Counts at Lake McConaughy between 2000–2001 and 2009–2010.

Habitats: Rivers, reservoirs, lakes and coastal shorelines are normally used by migrants.

Comments: This gull has at times been considered a race of the herring gull, and at other times the Iceland gull has been taxonomically expanded to racially include the Thayer's gull. Although named the Iceland gull, this species rarely breeds there, but does breed along the eastern and western coasts of Greenland. The world population of nominate *glaucoides* was estimated at about 40,000 pairs in the 1990's (B.O.N.A. 699).

Suggested Reading: Richards, 1990; B.O.N.A. 699; Howell and Dunn, 2007.

Lesser Black-backed Gull, *Larus fuscus*

A regular rare to occasional fall and early spring migrant in Nebraska and Kansas. Rare in winter, and a very rare or accidental vagrant in South Dakota.

Migration: By 2006 there were over 50 Nebraska records (*Nebraska Bird Review* 74:86), mostly occurring during spring and fall, with many February and March sightings. Sharpe, Silcock and Jorgensen (2001) noted that fall reports range from August 24 to December 19, and winter-to-spring reports extend from January 14 to April 19. There are only a few mid-winter records between January 3 and February 12. It has also been reported at least eight times in South Dakota, in February, April and November (Tallman, Swanson and Palmer, 2002). It has also been reported in at least 13 Kansas counties, from September to March (Thompson *et al.*, 2011). There were at least 30 spring Nebraska records as of 2006, and about 40 fall Nebraska records as of 2009 (*Nebraska Bird Review* 77:27).

Habitats: Rivers, reservoirs, lakes and coastal shorelines are normally used by migrants.

Comments: This gull species breeds coastally in northern Europe and Iceland. It is not yet known to breed in North America, but migrants regularly reach the Atlantic coast. It is similar in plumage to the great black-backed gull, but is about the same size as the herring gull.

Suggested Reading: Richards, 1990; Howell and Dunn, 2007.

Glaucous Gull, *Larus hyperboreus*

A rare overwintering migrant in Nebraska and Kansas; an accidental vagrant in South Dakota. Nebraska records are concentrated at Lake McConaughy, but county records include Douglas, Dawes, Harlan, Keith, Garden, Lancaster, Lincoln and Scotts Bluff counties. It has also been reported at least seven times in South Dakota, mostly in May and November (Tallman, Swanson and Palmer, 2002). Also reported from at least 14 mostly central and eastern Kansas counties, mostly between November and March (Thompson *et al.*, 2011).

Migration: Ten total spring Nebraska records range from January 24 to April 29, with a median of March 24. Six fall Nebraska records are from October 16–27. Sharpe, Silcock and Jorgensen (2001) noted that reports extend from November 16 to May 2. The birds are most frequent on large reservoirs from December to March.

Habitats: Rivers, reservoirs, lakes and coastal shorelines are normally used by migrants.

Comments: This close relative of the herring gull sometimes hybridizes with it, but breeds at generally higher latitudes, to extreme northern Canada and northern Greenland. Population estimates are highly tentative, but Alaska might have over 100,000 birds, and Canada over 70,000 (B.O.N.A. 573). It has a circumpolar breeding range, reaching the northernmost landmasses of the northern hemisphere. Unlike any of our other very large gulls, it has no black in the primary feathers at any age.

Suggested Reading: Richards, 1990; B.O.N.A. 573; Howell and Dunn, 2007.

Great Black-backed Gull, *Larus marinus*

A rare to extremely rare overwintering migrant in Nebraska and Kansas. County records for Nebraska include at least Dakota, Hamilton, Keith, and Lancaster counties. It also has been reported from seven central and eastern Kansas counties. Not yet reported for South Dakota..

Migration: By 2009 there were at least 21 total Nebraska records, six of which were midwinter records (*Nebraska Bird Review* 77:27). Sharpe, Silcock and Jorgensen (2001) noted that Nebraska reports extend from November 19 to April 13, with many January and February records. There is also one summer record, and it was observed four years during Christmas Bird Counts at Lake McConaughy between 2000–2001 and 2009–2010. Kansas reports are mostly from December to February (Thompson *et al.,* 2011).

Habitats: Rivers, reservoirs, lakes and coastal shorelines are normally used by migrants.

Comments: This is the largest of the gulls likely to appear in the Great Plains. It breeds along the northern cost of Eurasia, as well as in Iceland, Greenland and Newfoundland. Its range is slowly expanding both southward and northward, perhaps as a result of climate change. It is omnivorous, and powerful enough to kill rabbits. It also kills and eats the young of many seabirds, and consumes a great variety of marine life. The North American population was estimated at about 6,000 individuals in the 1990's (B.O.N.A. 330).

Suggested Reading: Richards, 1990; B.O.N.A. 330; Howell and Dunn, 2007.

ORDER PASSERIFORMES – PERCHING BIRDS

Family Parulidae – Wood Warblers

Northern Waterthrush, *Parksia noveboracensis*

An uncommon spring and fall migrant throughout the region, perhaps becoming rarer westwardly.

Migration: The range of 135 initial spring Nebraska sightings is from April 10 to May 27, with a median of May 7. Half of the records fall within the period May 2–11. Twenty-six final spring sightings are from May 3–21, with a median of May 14. Eight initial fall sightings are from August 10 to September 10, with a mean of August 29. Seven final fall sightings range from September 9 to October 12, with a mean of September 22. Sharpe, Silcock and Jorgensen (2001) noted that spring reports extend from April 20 to May 28, and fall reports from August 7 to September 30. There are also a few summer records.

Habitats: While in Nebraska this species is associated with deciduous forests or woodlands near streams.

Comments: Like ovenbirds, waterthrushes forage for invertebrates on the ground, often in wet locations. The song of this species also somewhat resembles the distinctive "*teacher*" song of the ovenbird, but does not rise in pitch and volume, but rather drops toward the end. In spite of a few summer records, there is no evidence of breeding in the region. National Breeding Bird Surveys between 1966 and 2009 indicate that the species underwent a statistically non-significant population increase (0.8 percent annually) during that period. Its North American population has been estimated at about 13 million birds (Rich *et al.*, 2004).

Very Rare, Vagrant, & Presumed Extinct Species

Family Anatidae – Ducks, Geese & Swans

Black-bellied Whistling-Duck. *Dendrocygna autumnalis*

Very rare migrant in Nebraska and Kansas. There have been 11 Nebraska records through 2010 (*Nebraska Bird Review* 78:133). There are also records for 13 Kansas counties (Thompson *et al.*, 2011), but none from South Dakota. This is a tropical species that has increasingly wandered northward in recent years, and has recently bred as far north as Oklahoma and Arkansas.

Bean Goose. *Anser fabalis*

Accidental vagrant in Nebraska. Photographed at De Soto National Wildlife Refuge, Washington County, from December 29, 1984 to January 10, 1985 (*Nebraska Bird Review* 53:3). Also reported April 4, 1998, Funk Waterfowl Production Area, Phelps County, (Brogie, 1999). Both were identified as belonging to the taiga or eastern Siberian race, *A. fabalis middendorfi.* This Asian species has not been reported from elsewhere in the region.

Emperor Goose. *Chen canagica*

Accidental vagrant in Nebraska. One was found dead at Harvard Waterfowl Production Area, Clay County, during the spring of 1997 (*Nebraska Bird Review*. 66:149, 153). This Alaskan and Asian species has not been reported from elsewhere in the region, and is rare anywhere south of Canada.

Brant. *Branta bernicla*

Extremely rare or an accidental vagrant in the three-state region. Besides some early Nebraska records for Buffalo and Hamilton counties; recent sight records are for Adams, Dawson, Kearney, Nemaha, and Webster counties. As of 2003 there were ten documented Nebraska records, four of which were of the Pacific race *nigricans* (*Nebraska Bird Review* 71:8, 64*)*. There are also records from 12 Kansas counties (Thompson *et al.*, 2011) and six South Dakota counties (Tallman, Swanson and Palmer, 2002). Brant winter commonly along the Atlantic and Pacific coasts, but rarely stray far inland.

Barnacle Goose. *Branta leucopsis*

Accidental European vagrant in Nebraska. A specimen was shot in Otoe County during November, 1968 (*Nebraska Bird Review* 37:2), and a probable escaped captive was seen in 1998 (*Nebraska Bird Review* 66:34). Barnacle geese have not been reported from elsewhere in the region, but are fairly regular visitors to the Atlantic coast.

Mute Swan. *Cygnus olor*

Hypothetical in Nebraska. Sightings from 1969 onward of this European swan were almost certainly of escaped captives, Mute swans are now feral in the Great Lakes and Atlantic Coast regions.

Garganey. *Anas querquedula*

An accidental Eurasian vagrant in Nebraska and Kansas. A male was seen on March 28, 1998 in Kearney County, Nebraska, and probably the same bird was later seen in Hall County, March 29–April 5, 1998 (*Nebraska Bird Review* 66:35, 149). There are also records from five Kansas counties (Thompson *et al.*, 2011).

Tufted Duck. *Aythya fuligula*

An accidental Eurasian vagrant in Nebraska. First observed at Lake Ogallala–Keystone, Keith County, in November 1999. Observed there for at least four successive winters, and reported on the Lake McConaughy/Ogallala Christmas Bird Counts on several occasions since 2000. There is one record for Sedgwick County, Kansas (Thompson *et al.,* 2011).

King Eider. *Somateria spectabilis*

An accidental coastal vagrant in Nebraska and Kansas. Photographed at DeSoto National Wildlife Refuge, 10–24 November, 1985 (Bray *et al.,* 1986). There is also one November, 1947, specimen record from Douglas County, Kansas (Thompson *et al.,* 2011).

Common Eider. *Somateria mollissima*

An accidental coastal vagrant in Nebraska and Kansas. There is a specimen record for a female of the Hudson Bay race *sedentaria* taken in early December of 1967 in Lincoln County, Nebraska (*Nebraska Bird Review* 37:38). There is also one November, 1893, specimen record from Douglas County, Kansas (Thompson *et al.,* 2011).

Harlequin Duck. *Histrionicus histrionicus*

An accidental vagrant in Nebraska and Kansas. Three early Nebraska records are from the Omaha area, one of which probably was from Burt County (Bruner, Wolcott and Swenk, 1904). There were three documented records for the state as of 2001, plus four undocumented ones (*Nebraska Bird Review* 69:164). There is one record for Wyandotte County, Kansas (Thompson *et al.,* 2011). The nearest breeding population occurs in the mountains of western Wyoming.

Family Gaviidae–Loons

Yellow-billed Loon. *Gavia adamsii*

An accidental coastal vagrant in Nebraska and Kansas. An adult was photographed on Branched Oak Lake, Lancaster County, November 17–21, 1996 (Brogie, 1997). An immature was seen on Lake McConaughy between August 8 and October 20, 1998 (Brogie, 1999), and one was seen at Lake Mc-Conaughy in September, 2003 (*Nebraska Bird Review* 71:150; 72:61. There are also records from five Kansas counties (Thompson *et al.*, 2011).

Family Phaethontidae–Tropicbirds

White-tailed Tropicbird. *Phaethon lepturus*

Hypothetical in Nebraska. A questionable sight record for this tropical species exists for Lincoln County (*Nebraska Bird Review* 41:59, 79). There are no other regional records.

Family Fregatidae –Frigate-birds

Magnificent Frigate-bird. *Fregata magnificens*

Hypothetical in Nebraska; accidental vagrant in Kansas. There is a sight record for this tropical species by Lawrence Bruner in Cuming County from the spring of 1884. There are also records from five Kansas counties, including one specimen record (Thompson *et al.*, 2011).

Family Pelecanidae–Pelicans

Brown Pelican. *Pelecanus occidentalis*

Very rare in the three-state region. Migrants of this coastal species have been seen in Nebraska on at least 13 occasions. Swenk (1934) summarized five early records, and since then the species has been reported in Lincoln County in 1937, in Cherry and Keya Paha counties in 1955, and in Custer County in 1977. There are also records for Cedar, Dakota, Dodge, Harlan, Knox and Platte counties, and for De Soto National Wildlife Refuge (*Nebraska Bird Review* 59:150; 70:98; 73:81; 77:52). There are also records from 16 Kansas counties (Thompson *et al.*, 2011) and three South Dakota counties (Tallman, Swanson and Palmer, 2002).

Family Anhingidae–Anhingas

Anhinga. *Anhinga anhinga*

An accidental vagrant in Nebraska and Kansas. Nebraska records include Buffalo County, September, 1913, Hamilton County, May, 1955, Greeley County, April, 1975, along the Platte River, October, 1976, Sarpy County, April, 1978, and Indian Cave State Park in 2005 (*Nebraska Bird Review* 73:50). Sharpe, Silcock and Jorgensen (2001) listed four spring reports, from April 8 to May 5, and two fall reports, for September 20 and late October. The few available seasonal records for this tropically oriented species range from April to October. There are also records from eight Kansas counties (Thompson *et al.,* 2011).

Family Ardeidae–Herons & Bitterns

Reddish Egret. *Egretta rufescens*

An accidental southern vagrant in Nebraska and Kansas. One photographic record exists for Lake McConaughy, during September and October, 2000 (*Nebraska Bird Review* 68:146). There are also records from two Kansas counties (Thompson *et al.,* 2011).

Family Ciconiidae - Storks

Wood Stork. *Mycteria americana*

An accidental southern vagrant in South Dakota, Nebraska and Kansas. Although there is a 1925 sight record (Sarpy County) the only specimen known is one obtained in Hamilton County in the 1880s (Bruner, Wolcott and Swenk, 1904). There is one 1964 South Dakota record (Tallman, Swanson and Palmer, 2002). There are also records from ten Kansas counties (Thompson *et al.,* 2011).

Family Threskiornithidae–Ibises & Spoonbills

White Ibis. *Eudocimus albus*

An accidental southern vagrant in South Dakota, Nebraska and Kansas. Observed over a period of several days in Rock County in August of 1963 *(Nebraska Bird Review* 32:12). Also reported July 5, 1999, at Kissinger Wildlife Management Area, Clay County (*Nebraska Bird Review* 67:88), and at Funk Waterfowl Production Area, Phelps County, on August 9, 2001 (*Nebraska Bird Review* 69:162; 71:98). There are also records from seven Kansas counties (Thompson *et al.,* 2011) and four South Dakota counties (Tallman, Swanson and Palmer, 2002).

Roseate Spoonbill. *Ajaia ajaja*

An accidental southern vagrant in Nebraska and Kansas. There is a Nebraska specimen record from Buffalo County, obtained in June of 1932. There is also a sight record of two seen near Hastings, Clay County, in August 1966 (*Nebraska Bird Review* 34:77), and a single individual was seen near Nebraska City on August 5 & 14, 1997 (*Nebraska Bird Review* 65:162). Sharpe, Silcock and Jorgensen (2001) listed four summer or fall reports, from June 5 to September 16. There are also records from eight Kansas counties (Thompson *et al.,* 2011).

Family Rallidae–Rails, Coots and Gallinules

Clapper Rail. *Rallus longirostris*

An accidental coastal vagrant in Nebraska. A single specimen record exists for Stapelton, Logan County; a bird captured in a trap January 30, 1951. It has not been reported from elsewhere in the region.

Purple Gallinule. *Porphyrula martinica*

An accidental southern vagrant in Nebraska; very rare in Kansas. One was observed in Cuming County in the summer of 1884 or 1885, and a second one was observed in Gage County on March 28, 1962 (*Nebraska Bird Review* 38:50). There is also a sighting from Adams County, May 2, 1946. There are records for at least 11 Kansas counties (Thompson *et al.*, 2011).

Family Gruidae–Cranes

Common Crane. *Grus grus*

An accidental Eurasian vagrant in Nebraska and Kansas. Apparently two different individuals were observed in Lincoln County during 1972, and in 1974 one was seen in Kearney County on March 16 and 25. An adult was reported from March 30–31, 1996 in Hall County (*Nebraska Bird Review* 64:80–82; Brogie, 1997), and during early March, 1999, in Kearney and Buffalo counties. A probable male with a sandhill crane mate and two apparently hybrid offspring were seen in Hall County in mid-March, 2000. Between 2007 and 2010 there were six reports of common cranes at locations between Buffalo and Garden counties, mostly in mid-March. Some of these reports may have been of the same bird (*Nebraska Bird Review* 78:47). There is also one Kansas sighting from Quivira N.W.R. (Thompson *et al.*, 2011).

Family Charadriidae–Plovers

Wilson's Plover. *Charadrius wilsonia*

Hypothetical in Kansas, with three sight records (Thompson, *et al.*, 2011). It has not been reported from elsewhere in the region.

Family Scolopacidae–Sandpipers and Phalaropes

Eskimo Curlew. *Numenius borealis*

Extirpated, and very probably extinct nationally. At one time this small curlew was a fairly common to abundant spring migrant in the Platte Valley of Nebraska, which was perhaps its most important spring stopover site in the Great Plains.

Migration: Ten spring Nebraska records are from March 22 to approximately May 25, with a median of April 12. There are no specific fall Nebraska records. Some birds perhaps migrated through the state in October, although the bulk of the population migrated from their breeding grounds to the Atlantic Coast, and then apparently flew nonstop south across the western Atlantic to the coast of northeastern South America, eventually wintering in southern South America. There are records from eight Kansas counties: Barton, Dickinson, Douglas, Ellis, Lyon, Russell, Sedgwick, and Woodson (Thompson *et al.,* 2011), and from five South Dakota counties: Brown, Charles Mix, Clay, Douglas and Yankton (Tallman, Swanson and Palmer, 2002).

Habitats: While in Nebraska this species settled in large flocks on newly plowed fields and dry, burnt-off prairies, where they foraged on grasshoppers and other insects, in a manner similar to buff-breasted sandpipers, black-bellied plovers and golden-plovers. They evidently concentrated in York, Fillmore and Hamilton counties, in flocks of up to several hundred birds. As the native prairies disappeared, the curlews increasingly used wheat fields and tame meadows.

Comments: The most recent accepted U.S. sight record for this species was a group of 23 observed in 1981 in Texas. There are also unconfirmed sight records from Canada in mid-May of 1996, but the last known specimen record was taken in Barbados in 1963. The last accepted sight record from Nebraska was a group of eight on April 8, 1926, near Hastings, and the last Kansas specimen record is from 1902. A Nebraska sight record of April 16, 1986 (*Nebraska Bird Review* 55:78) was not accepted by the N.O.U. Records Committee.

Sharp-tailed Sandpiper. *Calidris acuminata*

An accidental Eurasian vagrant in Nebraska, with three sight records of juveniles. The first was in Butler County, October 12, 1986 (*Nebraska Bird Review* 54:70, the second was one seen in Sheridan County, September 6, 1996 (*Nebraska Bird Review* 62:114), and the third was in Scotts Bluff County, September 18, 2002 (*Nebraska Bird Review 70:145).* It has not been reported from elsewhere in the region.

Curlew Sandpiper. *Calidris ferruginea*

An accidental Eurasian vagrant in Nebraska and Kansas. Seen at Funk Lagoon, Phelps County, Nebraska, July 19 and 21, 1997 (*Nebraska Bird Review* 66:3; 154). There are also records from three Kansas counties (Thompson *et al.,* 2011).

Ruff. *Philomachus pugnax*

An accidental Eurasian vagrant in the three-state region. Seven Nebraska records had accumulated as of 2005. Sharpe, Silcock and Jorgensen (2001) listed three documented spring reports, from April 9 to May 24, and two documented fall reports, from September 22 to September 27. There are also records from nine Kansas counties (Thompson *et al.,* 2011), and for three South Dakota counties (Tallman, Swanson and Palmer, 2002).

Red Phalarope. *Phalaropus fulicarius*

Very rare in the three-state region. The first Nebraska specimen was collected in Cherry County (*Nebraska Bird Review* 2:38). As of 2002 there have been at least nine fall Nebraska records, mostly of young birds (*Nebraska Bird Review* 70:14). Sharpe, Silcock and Jorgensen (2001) listed eight late summer to early fall reports, from August 1 to October 15. There are also records from 13 Kansas counties (Thompson *et al.,* 2011), and from five South Dakota counties (Tallman *et al,* 2002).

Family Stercorariidae - Jaegers

Pomarine Jaeger. *Stercorarius pomarinus*

Very rare in Nebraska; accidental vagrant in Kansas and South Dakota. Twelve state records have accumulated through 2005 (*Nebraska Bird Review* 73:141). Sharpe, Silcock and Jorgensen (2001) listed seven documented summer and fall reports, from June 30 to December 17. There are also records from eight Kansas counties (Thompson *et al.*, 2011) and three South Dakota counties (Tallman, Swanson and Palmer, 2002). This is the species of jaeger most often seen in the region, which has a circumpolar breeding distribution and normally migrates along the North American coastlines to wintering areas off the coasts of South America and other land masses.

Parasitic Jaeger. *Stercorarius parasiticus*

An accidental coastal vagrant in the three-state region. There were five state records through 2003. Sharpe, Silcock and Jorgensen (2001) listed two documented fall reports, for August 23–24 and October 5. There are also records from ten Kansas counties (Thompson *et al.*, 2011) and from one South Dakota county (Tallman, Swanson and Palmer, 2002). Like the other jaegers, this circumpolar species winters at sea off South America and elsewhere.

Long-tailed Jaeger. *Stercorarius longicaudus*

An accidental coastal vagrant in the three-state region. There is a specimen record from Lancaster County in 1952 (*Nebraska Bird Review* 21:2–3), and four more recent records, mostly of immatures (*Nebraska Bird Review* 69:170; 77:110–111). Sharpe, Silcock and Jorgensen (2001) listed two documented fall reports of young birds, for September 1 and October 3. There are also records from three Kansas counties (Thompson *et al.*, 2011) and two South Dakota counties (Tallman, Swanson and Palmer, 2002). This is a highly pelagic circumpolar species, usually wintering at sea, far off any coastline.

Family Laridae – Gulls and Terns

Black-legged Kittiwake. *Rissa tridactyla*

Very rare in South Dakota, Nebraska and Kansas. There is a single specimen record from Keith County, May, 1990 (*Nebraska Bird Review* 58:75). Several sight records exist from Lancaster County (*Nebraska Bird Review* 5:57, 49:42), the most recent from December 3, 1995. Also observed in Burt and Douglas counties during November, 1995 (Brogie, 1997). Sharpe, Silcock and Jorgensen (2001) listed seven spring reports (two documented), from April 20 to June 21, and 16 fall reports, from October 30 to December 23. Most Nebraska records are for November and December. Very rare in South Dakota, with about ten records (Tallman, Swanson and Palmer, 2001). This circumpolar arctic-breeding gull is highly pelagic, but strays often occur inland, and have been reported from at least 16 Kansas counties (Thompson *et al.*, 2001).

Black-headed Gull. *Chroicocephalus ridibundus*

An accidental Eurasian vagrant in Nebraska and Kansas. Three Nebraska reports (one of which was not accepted by the N.O.U Records Committee) existed by 2004, all from August to December. Reported at Walgren Lake, Sheridan County, August 12, 1979 (Rosche, 1982), and at Lake McConaughy in December, 2003 (*Nebraska Bird Review* 72:1). There are also records from eight Kansas counties (Thompson *et al.*, 2011). This Old World species breeds from Iceland east through Eurasia to the Kamchatka Peninsula, and is slowly invading northeastern North America.

Ross's Gull. *Rhodostethia rosea*

An accidental arctic vagrant in the three-state region. Observed at Sutherland Reservoir, Lincoln County, 17–23 Dec., 1992 (*Nebraska Bird Review* 61: 88–90), and at Branched Oak Lake, Lancaster County, November & December, 2010 (*Nebraska Bird Review* 79:13). There is one Kansas record (Thompson *et al.*, 2011) and some interstate South Dakota–Nebraska sightings at Gavin's Point Dam. Breeding of this small gull has been observed on the west coast of Hudson Bay, but North American nestings are very rare, and its primary nesting grounds are in northeastern Siberia.

Slaty-backed Gull. *Larus shistasagus*

Hypothetical in Nebraska. Reported January 22, 2000, at Harlan County Reservoir (*Nebraska Bird Review* 68: 19). This record was not accepted by the N.O.U. Records Committee. This herring gull-sized species is a Siberian gull that has not been reported from elsewhere in the region.

Glaucous-winged Gull. *Larus glaucescens*

An accidental Pacific-coast vagrant in Nebraska; hypothetical in Kansas. One was observed and photographed on April 12, 1995, at Lake McConaughy (*Nebraska Bird Review* 64:3–4). There is also a more recent December sighting from Nebraska, and sight records from Sedgwick and Riley counties, Kansas (Thompson *et al.,* 2011).

Arctic Tern. *Sterna paradisaea*

An accidental coastal vagrant in the three-state region. An adult was found September 20, 2000, at Lake Minatare (*Nebraska Bird Review* 68:158), one was seen at Lake Ogallala. December 21, 2003 (*Nebraska Bird Review* 72:61) and one was seen at Lake McConaughy June 11, 2006 (*Nebraska Bird Review* 74:87). There are also records from four Kansas counties (Thompson *et al.,* 2011) and from one South Dakota county (Tallman, Swanson and Palmer, 2002).

Royal Tern. *Thalasseus maximum*

An accidental coastal vagrant in Nebraska. There is a single specimen record from North Lake, York County (*Nebraska Bird Review 75:10*). It has not been reported from elsewhere in the region.

Family Alcidae - Auks

Ancient Murrelet. *Synthliboramphus antiquus*

An accidental Pacific-coast vagrant in Nebraska and South Dakota. There is a Nebraska single specimen record from Burt County (*Nebraska Bird Review* 1:14), and a specimen found Nov. 13, 1993 in Edmonds County, South Dakota. Not reported from elsewhere in the region, but there are two Colorado specimen records.

Long-billed Murrelet. *Brachyramphus perdix*

An accidental Asian vagrant in Kansas. Observed in Russell County on November 21-22, 1997 (Thompson *et al.*, 2011). It has not been reported from elsewhere in the region.

References

General and Regional References

Alderfer, J. (ed.). 2006. *National Geographic Complete Birds of North America.* Washington, D.C.: National Geographic Society.

Allen, D. L. 1967. *The Life of Prairies and Plains.* New York: McGraw-Hill.

Bellrose, F. C. 1980. *Ducks, Geese and Swans of North America.* 3rd ed. Harrisburg, PA: Stackpole Books.

Bent, A. C. 1942. Life histories of North American flycatchers, larks, swallows and their allies. *United States National Museum Bulletin* 179: 1–555.

_____. 1953. Life histories of North American wood warblers. *United States National Museum Bulletin* 203: 1–734.

_____. 1968. Life histories of North American cardinals, grosbeaks, buntings, towhees, finches, sparrows, and allies. In three parts. *United States National Museum Bulletin* 237:1–1889.

Brown, L., and D. Amadon, 1968. *Eagles, Hawks and Falcons of the World.* 2 vols. New York: McGraw-Hill.

American Ornithologists' Union (AOU). 1998. *The A.O.U. Checklist of North American Birds.* 7th ed. AOU, Washington, D.C. (annual supplements in *Auk* from 117: 847–856 to 127: 726–744).

Austin, J. E., and S. L. Richert. 2001. *A Comparative Review of Observational and Site Evaluation Data of Migrant Whooping Cranes in the United States, 1943–99.* Jamestown, ND: Northern Prairie Research Center.

Baldassarrie, G. A., and D. H. Fisher. 1984, Food habits of fall migrant shorebirds on the Texas High Plains. *J. of Field Ornithology* 55:220–229.

Berry, C. R. Jr. and D. G. Buechler. 1993. *Wetlands in the Northern Great Plains, A Guide to Values and Management.* Vermilion, SD: U.S. Fish and Wildlife Service and Agricultural Extension Service, S. D. State University. 13 pp.

Batt, B. D. 1996. Wetlands of the Great Plains region. Pp. 77–88, in *Prairie Conservation* (F. B. Samson & F. L. Knopf, eds.). Washington, D.C.: Island Press.

Batzer, D. B., and R. R. Shantz. 2006. *Wetlands.* Berkeley: Univ. of California Press.

Baumgartner, F. M., and M. Baumgartner. 1992. *Oklahoma Bird Life.* Norman: Univ. of Oklahoma Press.

Bolen, E. G., L. M. Smith, and H. L. Schramm, Jr. 1989. Playa lakes: Prairie wetlands of the southern High Plains. *Bioscience* 39:615–623.

Burton, J. F. 1995. Birds and Climate Change. London, UK: Christopher Helm.

Burt, W. 1994. *Shadowbirds: A Quest for Rails.* New York: Lyons & Burford.

Byrkjedal, I., and D. Thompson, 1998. *Tundra Plovers.* London: T&AD Poyser.

Center for Great Plains Studies (ed.). 1998. *Freshwater Functions and Values of Prairie Wetlands. Great Plains Research* (Special issue) 8(1): 1–208.

Curson. J., D. Quinn & D. Beadle. 1995. *Warblers of the Americas: An Identification Guide.* Boston: Houghton Mifflin.

Dahl, T. E. 1990. *Wetlands – Losses in the United States – 1780's to 1980's*. Washington, D.C.: U. S. Dept. of Interior, Fish & Wildlife Service.

Davis, C. A. 1996. Ecology of spring and fall migrant shorebirds on the playa lakes region of Texas. Ph.D. diss., Texas Tech. Univ., Lubbock, TX.

Dinsmore, J. J., T. H. Kent, D. Koenig, P. C. Peterson, and D. M. Roosa. 1984. *Iowa Birds*. Ames: Iowa State Univ. Press.

Eldridge, J. 1990. Aquatic invertebrates important for waterfowl production. Fish and Wildlife Leaflet 13.3.3. Washington, D.C.: U. S. Fish and Wildlife Service.

Eulis, N. H., Jr., D. A. Wrubleski, and D. M. Musket. 1999. Wetlands of the prairie pothole region: Invertebrate species composition, ecology and management. Pp. 471–514, in *Invertebrates in Freshwater Wetlands of North America* (D. P. Batzer, R. B. Rader and S. A. Wissinger, eds.). New York: John Wiley and Sons.

Fjeldsa, J. 2004. *The Grebes: Podicipedidae*. Oxford, UK: Oxford Univ. Press.

Hall, D. L., R. W. Sites, E. B. Fish, T. R. Mollhagen, D. L. Moorhead and M. R. Willig. 1999. Playas of the southern plains: The macroinvertebrate fauna. Pp. 635–665, in *Invertebrates in Freshwater Wetlands of North America; Ecology and Management* (D. P. Batzer, R. B. Rader and S. A. Wissinger, eds.). New York: John Wiley and Sons.

Hancock, J., and H. Elliott. 1978, *The Herons of the World*. New York: Harper and Row.

_____, and J. Kushlan. 1984. *The Heron Handbook*. New York: Harper & Row.

Harrington, B. A., and R. I. G. Morrison. 1979. Semipalmated sandpiper migration in North America. In *Shorebirds in Marine Environments* (F. A. Pitelka, ed.) Studies in Avian Biology 2: 83–100. Lawrence. KS: Cooper Ornithol. Soc., Allen Press.

Hayman, P., J. Marchant & T. Prater. 1986. *Shorebirds: An Identification Guide to the Waders of the World*. Boston: Houghton Mifflin.

Hitch, A. T., and P. L. Leeberg. 2007. Breeding distribution of North American bird species moving north as a result of climate change. Conservation Biology 21:54–59.

Howell, S. N. G., and J. Dunn. 2007. *Gulls of the Americas*. Boston: Houghton Mifflin.

Jehl, J. R., Jr. 1979. The autumnal migration of Baird's sandpiper. In *Shorebirds in Marine Environments* (F. A. Pitelka, ed.). Studies in Avian Biology 2:55–68. Lawrence. KS: Cooper Ornithol. Soc. & Allen Press.

Johnsgard, P. A. 1976. The grassy heartland. In *Our Continent: A Natural History of North America*. Washington, D. C.: National Geographic Society.

_____ . 1978. The ornithogeography of the Great Plains. *Prairie Naturalist*. 10:97–112.

_____ . 1973. *Grouse and Quails of North America*. Lincoln: Univ. of Nebraska Press.

_____ . 1975. *Waterfowl of North America*. Bloomington: Indiana University Press.

_____ . 1979. *Birds of the Great Plains: Breeding Species and their Distribution*. Lincoln : Univ. of Nebraska Press. Revised ed, with a Literature Supplement and revised maps. 2009. http://digitalcommons.unl.edu/bioscibirdsgreatplains/1/

_____ . 1981. *The Plovers, Sandpipers and Snipes of the World*. Lincoln: Univ. of Nebraska Press.

_____ . 1987. *Diving Birds of North America*. Lincoln: Univ. of Nebraska Press,

_____ . 1990. *Hawks, Eagles and Falcons of North America: Biology and Natural History*. Washington, D. C.: Smithsonian Institution Press,

_____. 1993. *Cormorants, Darters and Pelicans of the World*. Washington, D.C.: Smithsonian Inst. Press.

_____. 2001. *Prairie Birds: Fragile Splendor in the Great Plains*. Lawrence: Univ. Press of Kansas.

_____. 2008. *A Guide to the Natural History of the Central Platte Valley of Nebraska*. http://digitalcommons.unl.edu/biosciornithology/40

_____. 2009. *Birds of the Great Plains: Breeding Species and their Distribution*. Revised ed, with a literature supplement and revised maps. http://digitalcommons.unl.edu/bioscibirdsgreatplains/1/

_____. 2011. *The Sandhill and Whooping Cranes: Ancient Voices over the America's Wetlands*. Lincoln: Univ. of Nebraska Press.

_____, and M. Carbonell. 1996. *Ruddy Ducks and other Stifftails: Their Behavior and Biology*. Norman: Univ. of Oklahoma Press.

_____, and T. Shane . 2009. *Four Decades of Christmas Bird Counts in the Great Plains: Ornithological Evidence of a Changing Climate*. (Range maps by T. Shane) URL: http://digitalcommons.unl.edu/biosciornithology/46/

Kear, J. (ed.) 2005. *Ducks, Geese and Swans*. 2 vols. Oxford, UK: Oxford Univ. Press.

Keddy, P. A. 2010. *Wetland Ecology: Principles and Conservation*. Cambridge, UK: Cambridge Univ. Press

Küchler, A. W. 1964. *Potential natural vegetation of the conterminous United States*. American Geographical Society Special Publication no. 36. Washington, D. C.: American Geographical Society.

Kushlan, J. A., and J. A. Hancock. 2005. *Herons*. Oxford, UK: Oxford Univ. Press.

La Sorte, F. A., and F. R. Thompson III. 2007. Poleward shifts in winter ranges of North American birds. *Ecology* 88:1803–1812.

Leitch, J. A. and B. Hovde. 1996. Empirical valuation of prairie potholes: Five case studies. *Great Plains Research* 6:25–39.

Mills, A. M. 2005. Changes in the timing of spring and fall migration of North American migrant passerines during a period of global warming. *Ibis* 147:259–269.

Mitsch. W. J., and J. G. Gosselink. 2000. *Wetlands*. 3rd. ed. New York: Van Nostrand Reinhold Co.

_____, _____, Li Zhang, and C. J. Anderson. 2009. *Wetland Ecosystems*. New York, NY: John Wiley and Sons.

Morrison, E. I. G, 1984. Migration systems of some New World shorebirds. *Behav. Marine Animals* 6:125–202.

Morrison, R. I., R. E. Gill, Jr., B. A. Harrington, S. Skagen, G. W. Page, G. L. Gatto-Trevor, and S. M. Haig. 2001. Estimates of shorebird populations in North America. Ottawa: Canadian Wildlife Service, Environment Canada, *Occasional Paper* 104:1–64. (See also *J. Waterbird Soc.* 23:337–352. 2000.)

Murkin, H. R., A. G. van der Valk, and W. R. Clark. 2000. *Prairie Wetland Ecology, the Contribution of the Marsh Ecology Research Program*. Ames, IA: Iowa State Univ. Press. 413 pp.

Murphy-Klassen, H. M., T. J. Underwood, S. G. Sealy and A. A. Czernyi. 2005. Long-term trends in spring arrival dates of migrant birds at Delta Marsh, Manitoba, in relation to climate change. *Auk* 122:1130–1148.

Niering, W. A. 1991. *Wetlands of North America*. Charlottesville, VA: Thomasson-Grant.

Olsen, K. M., and K. Larson. 2004. *Gulls of North America, Europe and Asia*. Princeton, NJ: Princeton University Press.

Palmer, R. S. 1962. *Handbook of North American Birds*. Vol. 1. *Loons through Flamingos*. New Haven: Yale Univ. Press.

Peters, R. L, and T. E. Lovejoy (eds.).1992. *Global Warming and Biological Diversity*. New Haven, CT: Yale Univ. Press.

Poole, A. (ed.). varied dates. *The Birds of North America Online*, http://bna.birds. cornell.edu/ (Updated versions of the 715-part *The Birds of North America* monographs.)

Rich, T. C. *et al.* (eds.). 2004. *North American Landbird Conservation Plan*. Ithaca, NY: Partners in Flight and Cornell University Laboratory of Ornithology.

Richards, A. 1990. *Seabirds of the Northern Hemisphere*. New York: H. W. Smith.

Rising, J. 1996. *A Guide to the Identification and Natural History of the Sparrows of the United States*. New York: Academic Press.

Rose, P. M., & D. A. Scott. 1997. *Waterfowl Population Estimates* (2nd. ed.). Wageningen, Netherlands: Wetlands International.

Rose, P. M., & D. A. Scott. 2000. *Waterfowl Population Estimates* (3rd. ed.). Wageningen, Netherlands: Wetlands International,

Root, T. 1988. *Atlas of North American Winter Birds*. Chicago, IL: Univ. of Chicago Press.

Sauer, J. R., J. E. Hines, J. E. Fallon, K. L. Pardieck, D. J. Ziolkowski, Jr., and W. A. Link. 2011. *The North American Breeding Bird Survey, Results and Analysis 1966–2009. Version 3.23.2011* Laurel, MD: USGS Patuxent Wildlife Research Center.

Senner, S. E., and E. F. Martinez. 1982, A review of western sandpiper migration in interior North America. *Southwest. Nat.* 27:149–159,

Shaw, S., and C. G. Fredine. 1971. *Wetlands of the United States*. Washington, DC: Dept. of Interior, U. S. Fish and Wildlife Service, Circular 39.

Skagen, S. K., P. B. Sharpe, R. G. Waltermire, and M. B. Dillon. 1999. *Biogeographical Profiles of Shorebird Migration in Midcontinental North America*. U. S. Gov. Printing Office, Denver, CO: Biological Science Report USGS/BRD/BSR-2000-003. 167 pp.

Smith, L. M. 2003. *Playas of the Great Plains*. Austin, TX: Univ. of Texas Press. 257 pp.

Sowls, L. K. 1955. *Prairie Ducks: A Study of their Behavior, Ecology and Management*. Washington, D. C.: Wildlife Management Institute; Harrisburg, Pa.: Stackpole Company (reprinted 1978, Univ. of Nebraska Press).

Stavy, N. E., K. F. Dybala & M. A. Snyder. 2008. Climate models and ornithology. *Auk* 125:1–10.

Stewart. R. E. 1975. *Breeding Birds of North Dakota*. Fargo, ND: Tri-College Center for Regional Studies.

Steiert, J. 1985. *Playas: Jewels of the Plains*. Lubbock: Texas Tech. Univ. Press.

Tacha, T. C., and C. E. Braun (eds.). 1994. *Migratory Shore and Upland Game bird Management in North America*. Washington, D.C.: International Association of Fish & Wildlife Agencies.

Udvardy, M. D. F. 1958. Ecological and distributional analysis of North American birds. *Condor* 60:50–6.

U.S. Fish & Wildlife Service (U.S.F.W.S.). 2009a. *2009 Waterfowl Population Status*. Washington, D.C.: Administrative Report, U.S. Dept. of Interior. URL: http://www.fws.gov/migratorybirds/NewReportsPublications/PopulationStatus.html

_____. 2009b. *Migratory Bird Hunting Activity and Harvest during the 2007 and 2008 Hunting Seasons*. Compiled by R. V. Ratovich *et al*. Laurel, MD: U.S. Fish & Wildlife Service.

van der Valk, A. (ed.). 1989. *Northern Prairie Wetlands*. Ames, IA: Iowa State University Press. 400 pp

Weller, M. W. 1987. *Freshwater Marshes: Ecology and Wildlife Management*. Minneapolis, MN: Univ. of Minnesota Press.

Weller, M. W. and L. H. Fredrickson. 1973. Avian ecology of a managed glacial marsh. *Living Bird* 12:269–291,

Wetlands International. 2002. *Waterfowl Population Estimates–Third Edition*. Wageningen, The Netherlands: Wetlands International Global Series No. 12.

Wormworth, J., and K. Mullen. Undated. *Bird Species and Climate Change. The Global Status Report*. Version 1.0. A Report to World Wide Fund for Nature. 75 pp. http://assets.panda.org/downloads/birdsclimatereportfinal.pdf

State References

Kansas

Bowen, D. E. 1976. Coloniality, reproductive success and habitat interactions of upland sandpipers. *Bartramia longicauda*. Ph.D. diss., Kansas State Univ., Manhattan, KS.

Buchanan, R., ed. 1984. *Kansas Geology*. Lawrence, KS: Univ. Press of Kansas.

Busby, W. H. and J. L. Zimmerman. 2001. *Kansas Breeding Bird Atlas*. Lawrence, KS: Univ. Press of Kansas.

Cable, T. T., S. Seltman, and K. J. Cook. 1986. *Birds of Cimarron National Grassland*. Ft. Collins: U.S.D.A., Forest Service Gen. Tech. Rep. RM-GTR-281, Rocky Mtn. Forest & Range Experiment Station. 108 pp.

Chapman, S. S., *et al*. 2001. Ecoregions of Nebraska and Kansas (color poster and map). Reston, VA: U.S. Geological Survey.

Collins, J. T. 1973. *Natural Kansas*. Lawrence: Univ. Press of Kansas.

_____, S. I. Collins, J. Horak, D. Muhern, W. Busby, C. C. Freeman and G. Wallace. 1995. *An Illustrated Guide to Endangered and Threatened Species in Kansas*. Lawrence: Univ. Press of Kansas.

_____, B. Gress, G. Wiens, and S. I. Collins. 1991. *Kansas Wildlife*. Lawrence: Univ. Press of Kansas.

Gress, Bob, and P. Jensen, 2008. *Kansas Birds and Birding Hot Spots*. Lawrence, KS: Univ. Press of Kansas.

_____, and G. Potts. 1993. *Watching Kansas Wildlife: A Guide to 101 Sites*. Lawrence: Univ. Press of Kansas.

Hoffman, W. 1987. The birds of Cheyenne Bottoms. Pp. 327–362, In *Cheyenne Bottoms. An Environmental Assessment*. Kans. Biol. Survey and Kans. Geol. Survey.

Johnston, R. F. 1964. The breeding birds of Kansas. *University of Kansas Museum of Natural History Publications* 12:575–655.

_____. 1965. A directory to the birds of Kansas. *University of Kansas Museum of Natural History Miscellaneous Publications* 41:1–67.

Kansas Ornithological Society Bulletin. Published quarterly by the Kansas Ornithological Society.

Kuchler, A. W. 1974. A new vegetation map of Kansas. *Ecology* 55:586–604.

Parmelee, D. F., M. D. Schwilling and H. A. Stephens. 1969, Charadriiform birds of Cheyenne Bottoms. Part I. *Kans. Ornithol. Soc. Bull,* 20(2):9–13; Part II. 20(3):17–24.

_____, _____, & _____. 1970. Gruiform birds at Cheyenne Bottoms. *Kans. Ornithol. Soc. Bull.* 21(4): 25–27.

Rising, J. D. 1974. The status and faunal affinities of the summer birds of western Kansas. *University of Kansas Science Bulletin* 50:347–88.

Schwilling, M. 1985. Cheyenne Bottoms. *Kansas School Naturalist* 32(2): 3–15.

Thompson, M. C. & C. Ely. 1989, 1992. *Birds in Kansas.* 2 vols. Lawrence, KS: Univ. Press of Kansas.

_____, _____, B. Gress, C. Otte, S. T. Patti, D. Seibel & E. A. Young. 2011. *Birds of Kansas.* Lawrence, KS: Univ. Press of Kansas.

Tordoff, H. B. 1956. Check-list of the birds of Kansas. *University of Kansas Museum of Natural History Publications* 8:30–59.

Thompson M. C., and E. A. Young. 1993. Waterbirds of the Slate Creek Wetlands. Unpublished report to Kansas Dept. of Wildlife & Parks. 25 pp.

Young, E. A. 1993. A survey of the vertebrates of Slate Creek Salt Marsh, Sumner County, Kansas, with an emphasis on waterbirds. M.S. thesis, Fort Hayes State Univ., Hays, KS,

Zimmerman, J. L. 1990. *Cheyenne Bottoms: Wetland in Jeopardy.* Lawrence, KS: Univ. Press of Kansas.

_____. 1993. *The Birds of Konza: The Avian Ecology of the Tallgrass Prairie.* Lawrence, KS: Univ. Press of Kansas.

_____, and S. T. Patti. 1988. *A Guide to Bird-finding in Kansas and Western Missouri. Lawrence:* Univ. Press of Kansas.

Nebraska

Bleed, A., and C. Flowerday (eds.). 1989. *An Atlas of the Sand Hills* (A. Bleed and Resource Atlas No. 5. Lincoln, NE: Conservation and Survey Division. Univ. of Nebraska–Lincoln. 238 pp.

Bray, T. E., B K. Padelford and W. R. Silcock. 1986. *The Birds of Nebraska: A Critically Evaluated List.* Bellevue: Published by the authors. 109 pp.

Brogie, M. A. 1997. 1996 (Ninth) Report of the NOU Records Committee. *Nebraska Bird Review* 65:115–125.

Brogie, M. A., & M. J. Mossman. 1983. Spring and summer birds of the Niobrara Valley Preserve area, Nebraska. *Nebraska Bird Review* 51: 44–51.

Brown, C. R, M. B. Brown, P. A. Johnsgard, J. Kren & W. C. Scharf. 1996. Birds of the Cedar Point Biological Station area, Keith and Garden Counties, Nebraska: Seasonal occurrence and breeding data. *Transactions of the Nebraska Academy of Sciences* 29: 91–108.

_____, and _____. 2001. *Birds of the Cedar Point Biological Station.* Lincoln, NE: Occasional Papers of the Cedar Point Biological Station No. 1. 36 pp.

Brown, M., S. Dinsmore, and J. Jorgensen. 2012. *The Birds of Southwestern Nebraska.* Lincoln, NE: Division of Conservation and Survey, Univ. of Nebraska School of Natural Resources.

Bruner, L. R., H. Wolcott & M. H. Swenk. 1904. *A Preliminary Review of the Birds of Nebraska.* Omaha: Klopp & Bartlett. 116 pp.

Cariveau, A. B., L. A. Johnson, and R. A. Sparks. 2007. *Biological Inventory and Evaluation of Conservation Strategies in Southwestern Playa Wetlands.*. Brighton, CO: Rocky Mountain Bird Observatory, Report to Nebraska Game & Parks Comm. and Playa Lakes Joint Venture. 39 pp. URL: http://www.rmbo.org/dataentry/postingArticle/dataBox/RNBO-E

_____, and D. Pavlacky. 2009. *Biological Inventory and Evaluation of Conservation Strategies in Southwestern Playa Wetlands.* Final Report to Nebraska Game and Parks Comm. and Playa Lakes Joint Venture. Brighton, CO: Rocky Mountain Bird Observatory. 74 pp. (Gov. Doc. No. G100 B138-2009)

Chapman, S. S., *et al.* 2001. Ecoregions of Nebraska and Kansas (color poster and map). Reston, VA: U.S. Geological Survey.

Cunningham, D., and T. Krueger (eds.). 1996. Weather and climate of Nebraska. *Nebraskaland Magazine* 74(1): 1–138.

Currier, P. J., G. R. Lingle, and J. G. VanDerwalker. 1985. *Migratory Bird Habitat on the Platte and North Platte Rivers in Nebraska.* Grand Island, NE: Whooping Crane Habitat Maintenance Trust. 177 pp.

Ducey, J. E. 1988. *Nebraska Birds: Breeding Status and Distribution.* Omaha: Simmons-Boardman Books.

_____. 1989. Birds of the Niobrara River Valley. *Trans. Nebraska Acad. Sci.* 17:37–60.

Evans, R. D. and C. W. Wolfe, Jr. 1967. Waterfowl production in the Rainwater Basin area of Nebraska. *Journal of Wildlife Management* 33: 788–794.

Faanes, C. E., and G. R. Lingle. 1995. Breeding birds of the Platte Valley of Nebraska. Northern Prairie Wildlife Research Center Home Page, Jamestown, ND. URL= http://www.npwrc.usgs.gov/resources/distr/birds/platte/platte (version 16JUL97)

Farrar, J. 1982. The Rainwater Basin: Nebraska's vanishing wetlands. *Nebraskaland* 60(3): 18–41.

_____. 1984. Trumpeters. *Nebraskaland* 62(2):22-29.

_____. 1991. Marsh birds. *Nebraskaland* 69:(4):8-21.

_____. 1991. Nebraska salt marshes: Last of the least. *Nebraskaland* 69(6): 18–41.

_____. 1996. Nebraska's Rainwater Basin. *Nebraskaland* 74(2): 18–35.

_____. 1997. Cormorants. *Nebraskaland* 75(4):18-27.

_____. 2004. Birding Nebraska. *Nebraskaland Magazine* 82(1):1–178.

Gersib, R. A., B. Elder, K. F. Dinan and T. H. Hupf. 1989. *Waterfowl values by wetland type within Rainwater Basin wetlands with special emphasis on activity time budget and census data.* Grand Island, NE: Nebraska Game and Parks Commission and U.S. Fish and Wildlife Service. 105 pp.

Gordon, C. C., L. D. Flake, and K. F. Higgins. 1990. Aquatic invertebrates in the Rainwater Basin area of Nebraska. *Prairie Nat.* 22:191–200.

Johnsgard, P. A. 1979. The breeding birds of Nebraska. *Nebraska Bird Review* 47: 3–14.

_____. 1995. *This Fragile Land: A Natural History of the Nebraska Sandhills.* Lincoln, NE: Univ. of Nebraska Press.

_____. 1996. The cranes of Nebraska. *Museum Notes* (U. of Nebraska State Museum) 93:1-4.

_____. 1998. A half-century of winter bird surveys at Lincoln and Scottsbluff, Nebraska. *Nebraska Bird Review* 66:74–84. URL: http://digitalcommons.unl.edu/cgi/viewcontent.cgi?article=1004

———. 2001. *The Nature of Nebraska: Ecology and Biodiversity.* Lincoln, NE: Univ. of Nebraska Press.

———. 2007. *The Birds of Nebraska.* Revised edition. http://digitalcommons.unl. edu/biosciornithology/38

———. 2008. *A Guide to the Natural History of the Central Platte Valley of Nebraska.* http://digitalcommons.unl.edu/biosciornithology/40

———. 2011. *A Nebraska Bird-finding Guide.* Lincoln, NE: Zea Books. URL: http:// digitalcommons.unl.edu/biosciornithology/51

———. 2012. *The Wetlands of Nebraska.* Lincoln, NE: Division of Conservation and Survey, Univ. of Nebraska School of Natural Resources.

———, and J. Dinan. 2006 - Habitat associations of Nebraska Birds. *Nebraska Bird Review,* 73:20-25. http://www.nebraskabirds.org/Resources/resources.htm

———, and K. Gill. 2011. Sandhill cranes: Nebraska's avian ambassadors at large. *Prairie Fire,* March. Pp. 14. 15, 20. http://www.prairiefirenewspaper. com/2011/02/sandhill-cranes-our-avian-ambassadors-at-large

Jones, S. 2000. *The Last Prairie.* Camden, ME: Ragged Mountain Press.

Jorgensen. J. 2004. *An Overview of Shorebird Migration in the Eastern Rainwater Basin.* Occasional Paper No. 8, Nebr. Ornithol. Union, Lincoln, NE. 68 pp.

Krapu, G. L. (ed.). 1996. *The Platte River Ecology Study:* Special Research Report. Jamestown, ND: Northern Prairie Wildlife Research Station. 186 pp. URL: http://www.npsc.gov/resource/othrdata/platteco/platteco.htm

Labedz, T. 1989. Birds. Pp. 161–180, in *An Atlas of the Sand Hills* (A. Bleed and C. Flowerday, eds.). Resource Atlas No. 5. Lincoln, NE: Conservation and Survey Division. Univ. of Nebraska–Lincoln. 238 pp.

Lingle, G. R. 1994. *Birding Crane River: Nebraska's Platte.* Grand Island, NE: Harrier Pub. Co. 122 pp.

———, and G. L. Krapu. 1986. Winter ecology of bald eagles in south-central Nebraska. *Prairie Naturalist* 18:65–78.

LaGrange, T. G. 2005. *Guide to Nebraska Wetlands and their Conservation Needs.* Lincoln, NE: Nebraska Game and Parks Commission. 2nd ed. 57 pp. URL: www. nebraskawetlands.com

McCarraher, D. B. 1977. *Nebraska's Sandhills Lakes.* Lincoln: Nebraska Game and Parks Comm. 67 pp.

McMurtry, M. S., R. Craig, and G. Schildmann. 1972. Nebraska wetland survey. Lincoln: Nebraska Game and Parks Comm. 78 pp.

McCrae, D. B. 1972. The small playa lakes of Nebraska: Their ecology, fisheries and biological potential. Pp. 15–23, in *Playa Lakes Symposium,* 29–30 October, 1970. Lubbock, TX: Internat. Center for Arid and Semi-Arid Land Studies and Dept. of Geosciences, Texas Tech Univ. Publ. No. 4.

Mollhoff, W. J. 2001. *The Nebraska Breeding Bird Atlas.* Nebraska Game & Parks Comm. Lincoln.

Moulton, M. P. 1972. The small playa lakes of Nebraska: Their ecology, fisheries and biological potential. Pp. 15-23, in *Playa Lakes Symposium Transactions,* Internat. Center for Arid and Semi-Arid Land Studies. Texas Tech Univ. Lubbock.

Nebraska Ornithologists' Union (N.O.U.).1987–2010. Reports of the N.O.U. Records Committee. *Nebraska Bird Review* 55:79–85. 57:42–47; 58:90–97; 60:150–155; 64:38–42; 65:115–126; 66:147–159; 67:141–152; 69:85–91; 70:84–90; 71:97–102; 71:136–142; 72:59–65; 73:78–84; 74:69–74; 75:86–94; 76:111–119; 77:160–168.

_____. 2009. The official list of the birds of Nebraska. *Nebraska Bird Review* 77:112–131 (see also 56:86–96, 65:3–16, 72:108–126 and 73:84).

Nebraska Bird Review. Published quarterly by the Nebraska Ornithologists' Union. 78 vols. through 2010.

Novacek, J. M. 1989. The water and wetland resources of the Nebraska Sandhills. Pp. 340–384, in *Northern Prairie Wetlands* (A. van der Valk, ed.). Ames, IA: Iowa State University Press. 400 pp.

Oberholser, H. C, and W. L. McAtee. 1920. Waterfowl and their food plants in the Sandhill region of Nebraska. Washington, D.C.: *U.S. Dept. of Agriculture Bulletin* 794:1–79.

Panella, M. J. 2010. *Nebraska's At-risk Wildlife.* Lincoln: Nebraska Game and Parks Commission. 196 pp.

Poor, J. P. 1999. The value of additional Central Flyway wetlands: The case of Nebraska's Rainwater Basin wetlands. *J. of Agricultural and Resource Economics* 24:254–265.

Rosche, R. C. 1977. *Check-list of birds of northwestern Nebraska and southwestern South Dakota.* Crawford, Neb.: R. C. Rosche.

Sharpe, R. S., W. R. Silcock, and J. G. Jorgensen. 2001. *Birds of Nebraska: Their Distribution and Temporal Occurrence.* Lincoln, NE: Univ. of Nebraska Press

South Dakota

Bryce, S., J. M. Omernik, D. E. Pater, M. Ulmer, J, Schaar, J, Freeouf, R. Johnson, P. Kuck, and S. H. Azevedo. 1998. Ecoregions of North Dakota and South Dakota. URL: http://www.npwrc.usgs.gov/resource/habitat/ndsdeco/index.htm

Harrell, B. E., ed. 1978. *The Birds of South Dakota: An annotated check-list.* Vermilion: South Dakota Ornithologists' Union and W. H. Over Dakota Museum.

Hubbard, D. E. 1989. *Wetland Values in the Prairie Pothole Region of Minnesota and the Dakotas.* Brookings, SD: U.S. Fish and Wildlife Service, Cooperative Research Unit, Biological Report 88(43).

Johnson, R. R., K. F. Higgins, M. L. Kjellsen & C. R. Elliott. 1997. *Eastern South Dakota Wetlands.* Brookings: South Dakota State Univ. 28 pp. URL: http://www.npwrc.usgs.gov/resource/wetlands/eastwet/index.htm

Johnson, R. R. and K. F. Higgins. 1997. *Wetland Resources of Eastern South Dakota.* Brookings: South Dakota State Univ. 120 pp. URL: http://www.npwrc.usgs.gov/resource/wetlands/sdwet/index.htm

Kantrud, H. A., G. L. Krapu, and G. A. Swanson. 1989. *Prairie Basin Wetlands of the Dakotas: A Community Profile.* U. S. Fish and Wildlife Service, Biological Report 85(7.28). 111 pp. URL: http://www.npwrc.usgs.gov/resource/wetlands/basin-wet/index.htm

Lohoefener, R., and C. A. Ely. 1978. The nesting birds of Lacreek National Wildlife Refuge. *South Dakota Bird Notes* 30:14–30.

McLeod, Scott J., and Kenneth F. Higgins. 1998. *Waterfowl and habitat changes after 40 years on the Waubay study area.* South Dakota Agricultural Experiment Station Bulletin 728. South Dakota State University, Brookings, South Dakota. 40 pp. URL: http://www.npwrc.usgs.gov/resource/birds/waubay/index.htm

Meeks, William A., and Kenneth F. Higgins. 1998. Nongame birds, small mammals, herptiles, fishes: Sand Lake National Wildlife Refuge, 1995-1996. South Dakota Agricultural Experiment Station Bulletin 729. South Dakota State University,

Brookings, South Dakota. 28 pp. URL: http://www.npwrc.usgs.gov/resource/wildlife/sandlake/index.htm

Peterson, R. A. 1995. *The South Dakota Breeding Bird Atlas.* So. Dakota Ornithologists' Union, Aberdeen, SD. URL: http://www.npwrc.usgs.gov/resource/birds/sdatlas/index.htm

Pettingill, O. S., Jr., and Whitney, N. R., Jr. 1965. *Birds of the Black Hills.* Ithaca, N.Y.: Cornell Laboratory of Ornithology Special Publication no. I.

Rosche, R. C. 1977. *Check-list of birds of northwestern Nebraska and southwestern South Dakota.* Crawford, Neb.: R. C. Rosche.

Schwalbach, M., G. Vandel and K. Higgins. 1996. *Status, Distribution and Production of the Interior Least Tern and Piping Plover along the Mainstem Missouri River in South Dakota.* Pierre, SD: South Dakota Game. Fish and Parks Report 86:10.

South Dakota Bird Notes. Published quarterly by South Dakota Ornithologists' Union.

South Dakota Ornithologists' Union (S.D.O.U). 1991. *The Birds of South Dakota.* Aberdeen: Northern State Univ. Press.

Tallman, D. A., D. L. Swanson, and J. S. Palmer. 2002. *Birds of South Dakota.* South Dakota Ornithologists' Union, Aberdeen.

Birds of North America (B.O.N.A.) Species Monographs

(Published 1990 *et seq.* by The Birds of North America, Inc. Philadelphia, PA: The Academy of Natural Sciences, and Washington, DC: the American Ornithologists' Union. Updated electronic versions for many species are now available online at: http://bna.birds.cornell.edu/).

Fulvous Whistling-Duck. W. L. Hohman and S. A. Lee. 2001. B.O.N.A. 562. 24 pp.

Black-bellied Whistling-Duck. J. D. James and J. E. Thompson. 2001. B.O.N.A. 578. 20 pp.

Greater White-fronted Goose. C. R. Elly & A. X. Dzubin. 1994. B.O.N.A. 131. 32 pp.

Snow Goose. T. B. Mowbray, F. Cooke, & B. Ganter. 2000. B.O.N.A. 514. 40 pp.

Ross's Goose. John P. Ryder and Ray T. Alisauskas. 1995. B.O.N.A. 162. 28 pp.

Canada Goose. T. B. Mowbray, C. R. Ely, J. S. Sedinger, and R. E. Trost. 2003. B.O.N.A. 682. 44 pp

Trumpeter Swan. Carl D. Mitchell. 1994. B.O.N.A. 105. 24 pp.

Tundra Swan. R. J. Limpert and S. L. Earnst. 1994. B.O.N.A. 89. 20 pp.

Wood Duck. Gary R. Hepp and Frank C. Bellrose. B.O.N.A. 169. 24 pp.

Gadwall. C. R. Leschack, S. K. McKnight, & G. R. Hepp. 1997. B.O.N.A. 283. 28 pp.

American Wigeon. Thomas Mowbray. 1999. B.O.N.A. 401. 32 pp.

American Black Duck. J. Longcore, D. McAuley, G. Hepp, & J. Rhymer. 2000. B.O.N.A. 481. 36 pp.

Mallard. Nancy Drilling, R. Titman, and F. McKinney. 2002. B.O.N.A. 658. 44 pp.

Mottled Duck. T. E. Moorman and P. N. Gray. 1994. B.O.N.A. 81. 20 pp.

Blue-winged Teal. R. C. Rohwer, W. P. Johnson, and E. R. Loos. 2002. B.O.N.A. 625. 36 pp.

Cinnamon Teal. James H. Gammonley. 1996. B.O.N.A. 209. 20 pp.

Northern Shoveler. Paul J. Dubowy. 1996. B.O.N.A. 217. 24 pp.
Northern Pintail. Jane E. Austin and Michael R. Miller. 1995. B.O.N.A. 163. 32 pp.
Green-winged Teal. Kevin Johnson. 1995. B.O.N.A. 193. 20 pp.
Canvasback. T. B. Mowbray. 2002. B.O.N.A. 659. 40 pp.
Redhead. M. C. Woodin and T. C. Michot. 2003. B.O.N.A. 695. 40 pp.
Ring-necked Duck. William L. Hohman & Robert T. Eberhardt. 1998. B.O.N.A. 329.
    32 pp.
Greater Scaup. B. Kessel, D. A. Rocque, and J. S. Barclay. 2002. B.O.N.A. 650. 23 pp.
Lesser Scaup. James E. Austin, C. M. Custer, & A. D. Afton. 1998. B.O.N.A. 338. 32
    pp.
Harlequin Duck. Gregory J. Robertson and R. Ian Goudie. 1999. B.O.N.A. 466. 32 pp.
Surf Scoter. Jean-Pierre L. Savard, Daniel Bordage, & A. Reed. 1998. B.O.N.A. 363.
    28 pp.
White-winged Scoter. Patrick W. Brown & L. H. Fredrickson. 1997. B.O.N.A. 274. 28
    pp.
Black Scoter. Daniel Bordage & Jean-Pierre L. Savard. 1995. B.O.N.A. 177. 20 pp.
Long-tailed Duck. G. J. Robertson and J. P. L. Savard. 2002. B.O.N.A. 651. 28 pp.
Bufflehead. Gilles Gauthier. 1993. B.O.N.A. 67. 24 pp.
Common Goldeneye. J. M. Eadie, M. L. Mallory, and H. G. Lumsden. 1995. B.O.N.A.
    170. 32 pp.
Barrow's Goldeneye. J. M. Eadie, J-P. L. Savard, and M. L. Mallory. 2000. B.O.N.A.
    548. 32 pp.
Hooded Merganser. B. D. & K. M. Dugger, and L. H. Fredrickson. 1994. B.O.N.A. 98.
    24 pp.
Common Merganser. Mark Mallory and Karen Metz. 1999. B.O.N.A. 442. 28 pp.
Red-breasted Merganser. Rodger D. Titman. 1999. B.O.N.A. 443. 24 pp.
Ruddy Duck. R. B. Brua. 2003. B.O.N.A. 696. 32 pp.
Red-throated Loon. J. F. Barr, C. Eberl, & J. W. McIntyre. 2000. B.O.N.A. 513. 28 pp.
Pacific Loon/Arctic Loon. R. W. Russell. 2002. B.O.N.A. 657. 40 pp.
Common Loon. Judith W. McIntyre & Jack F. Barr. 1997. B.O.N.A. 313. 32 pp.
Pied-billed Grebe. Martin J. Muller and Robert W. Storer. 1999. B.O.N.A. 410. 32 pp.
Horned Grebe. Stephen J. Stedman. 2000. B.O.N.A. 505. 28 pp.
Red-necked Grebe. Bonnie E. Stout and Gary L. Nuechterlein. 1999. B.O.N.A. 465.
    32 pp.
Eared Grebe. S. A. Cullen, J. R. Jehl Jr., and G. L. Nuechterlein. 1999. B.O.N.A. 433.
    28 pp.
Western Grebe/Clark's Grebe. R. W. Storer & G. L. Nuechterlein. 1992. B.O.N.A. 26.
    24 pp.
Neotropic Cormorant. R. C. Telfair & M. L. Morrison. 1995. B.O.N.A. 137. 24 pp.
Double-crested Cormorant. Jeremy J. Hatch and D. V. Weseloh. 1999. B.O.N.A. 441.
    36 pp.
American White Pelican. Roger M. Evans and Fritz L. Knopf. 1993. B.O.N.A. 57. 24
    pp.
American Bittern. J. P. Gibbs, S. Melvin, and F. A. Reid. 1992. B.O.N.A. 18. 12 pp.
Least Bittern. J. P. Gibbs, F. A. Reid, and S. M. Melvin. 1992. B.O.N.A. 17. 12 pp.
Great Blue Heron. Robert W. Butler. 1992. B.O.N.A. 25. 20 pp.
Great Egret. D. A. McCrimmon, Jr., J. C. Ogden, and G. T. Bancroft. 2001. B.O.N.A.
    570. 32 pp.

Snowy Egret. Katharine C. Parsons and Terry L. Master. 2000. B.O.N.A. 489. 24 pp.
Little Blue Heron. James A. Rodgers, Jr. and Henry T. Smith. 1995. B.O.N.A. 145. 32 pp.
Tricolored Heron. Peter C. Frederick. 1997. B.O.N.A. 306. 28 pp.
Cattle Egret. Raymond C. Telfair, II. 1994. B.O.N.A. 113. 32 pp.
Green Heron. W. E. Davis, Jr. & J. A. Kushlan. 1994. B.O.N.A. 129. 24 pp.
Black-crowned Night-Heron. William E. Davis, Jr. 1993. B.O.N.A. 74. 20 pp.
Yellow-crowned Night-Heron. Bryan D. Watts. 1995. B.O.N.A. 161. 24 pp.
White Ibis. James A. Kushlan and Keith L. Bildstein. 1992. B.O.N.A. 9. 20 pp.
Glossy Ibis. William E. Davis, Jr. and John Kricher. 2000. B.O.N.A. 545. 20 pp.
White-faced Ibis. Ronald R. Ryder & David E. Manry. 1994. B.O.N.A. 130. 24 pp.
Osprey. A. F. Poole, R. O. Bierregaard, and M. S. Martell. 2003. B.O.N.A. 683. 44 pp.
Bald Eagle. David A. Buehler. 2000. B.O.N.A. 506. 40 pp.
Northern Harrier. R. Bruce MacWhirter & Keith L. Bildstein. 1996. B.O.N.A. 210. 32 pp.
Yellow Rail. Theodore A. Bookhout. 1995. B.O.N.A. 139. 16 pp.
Black Rail. W. R. Eddleman, R. E. Flores, & M. L. Legare. 1994. B.O.N.A. 123. 20 pp.
King Rail. Brooke Meanley. 1992. B.O.N.A. 3. 12 pp.
Virginia Rail. Courtney J. Conway. 1995. B.O.N.A. 173. 20 pp.
Sora. Scott M. Melvin and James P. Gibbs. 1996. B.O.N.A. 250. 20 pp.
Purple Gallinule. R. L. West and G. K. Hess. 2002. B.O.N.A. 626. 28 pp.
Common Moorhen. B. K. Bannor and E. Kiviat. 2003. B.O.N.A. 685. 28 pp.
American Coot. I. L. Brisbin, Jr., H. D. Pratt, and T. B. Mobray. 2003. B.O.N.A. 697. 44 pp.
Sandhill Crane. T. C. Tacha, S. A. Nesbitt, and P. A. Vohs. 1992. B.O.N.A. 31. 24 pp.
Whooping Crane. James C. Lewis. 1995. B.O.N.A. 153. 28 pp.
Black-bellied Plover. D. R. Paulson. 1995. B.O.N.A. 186. 28 pp.
American/Pacific Golden-Plover. O. W. Johnson & P. G. Connors. 1996. B.O.N.A. 201-202. 40 pp.
Snowy Plover. G. W. Page, J. S. Warriner, J. C. Warriner, & P. W. C. Paton. 1995. B.O.N.A. 154. 24 pp.
Semipalmated Plover. Erica Nol and Michele S. Blanken. 1999. B.O.N.A. 444. 24 pp.
Piping Plover. Susan M. Haig. 1992. B.O.N.A. 2. 18 pp.
Killdeer. Bette J. S. Jackson & Jerome A. Jackson. 2000. B.O.N.A. 517. 28 pp.
Black-necked Stilt. J. A. Robinson, J. M. Reed, J. P. Skorupa, & L. W. Oring. 1999. B.O.N.A. 449. 32 pp.
American Avocet. J. A. Robinson, L. W. Oring, J. P. Skorupa, & R. Boettcher. 1997. B.O.N.A. 275. 32 pp.
Spotted Sandpiper. L. W. Oring, E. M. Gray, & J. M. Reed. 1997. B.O.N.A. 289. 32 pp.
Solitary Sandpiper. William Moskoff. 1995. B.O.N.A. 156. 16 pp.
Greater Yellowlegs. Chris S. Elphick & T. Lee Tibbitts. 1998. B.O.N.A. 355. 24 pp.
Willet. P. E. Lowther, H. D. Douglas III, and C. L. Gratto-Trevor. 2001. B.O.N.A. 579. 32 pp.
Lesser Yellowlegs. T. Lee Tibbitts and William Moskoff. 1999. B.O.N.A. 427. 28 pp.
Upland Sandpiper. C. Stuart Houston and Daniel E. Bowen, Jr. 2001. B.O.N.A. 580. 32 pp.
Eskimo Curlew. R. E. Gill, Jr., P. Canevari, & E. H. Iversen. 1998. B.O.N.A. 347. 28 pp.

Whimbrel. Margaret A. Skeel and Elizabeth P. Mallory. 1996. B.O.N.A. 219. 28 pp.
Long-billed Curlew. B. D. Dugger and K. M. Dugger. 2002. B.O.N.A. 628. 28 pp.
Hudsonian Godwit. C. S. Elphick and J. Klima. 2002. B.O.N.A. 629. 32 pp.
Marbled Godwit. Cheri L. Gratto-Trevor. 2000. B.O.N.A. 492. 24 pp.
Ruddy Turnstone. David N. Nettleship. 2000. B.O.N.A. 537. 32 pp.
Red Knot. Brian A. Harrington. 2001. B.O.N.A. 563. 32 pp.
Sanderling. B. MacWhirter, P. Austin-Smith, Jr., & D. Kroodsma. 2002. B.O.N.A. 653. 28 pp.
Semipalmated Sandpiper. Cheri L. Gratto-Trevor. 1992. B.O.N.A. 6. 20 pp.
Western Sandpiper. W. Herbert Wilson. 1994. B.O.N.A. 90. 20 pp.
Least Sandpiper. John M. Cooper. 1994. B.O.N.A. 115. 28 pp.
White-rumped Sandpiper. David F. Parmelee. 1992. B.O.N.A. 29. 16 pp.
Baird's Sandpiper. W. Moskoff. 2002. B.O.N.A. 661. 20 pp.
Pectoral Sandpiper. Richard T. Holmes and Frank A. Pitelka. 1998. B.O.N.A. 348. 24 pp.
Dunlin. Nils D. Warnock & Robert E. Gill. 1996. B.O.N.A. 203. 24 pp.
Stilt Sandpiper. Joanna Klima & Joseph R. Jehl, Jr. 1998. B.O.N.A. 341. 20 pp.
Buff-breasted Sandpiper. R. B. Lanctot and C. D. Laredo. 1994. B.O.N.A. 91. 20 pp.
Short-billed Dowitcher. J. R. Jehl, Jr., J. Klima, and R. Harris. 2001. B.O.N.A. 564. 28 pp.
Long-billed Dowitcher. John Y. Takekawa and Nils Warnock. 2000. B.O.N.A. 493. 20 pp.
Wilson's (Common) Snipe. Helmut Mueller. 1999. B.O.N.A. 417. 20 pp.
American Woodcock. D. M. Keppie and R. M. Whiting, Jr. 1994. B.O.N.A. 100. 28 pp.
Wilson's Phalarope. M.A. Colwell and J.R. Jehl, Jr. 1994. B.O.N.A. 83. 20 pp.
Red-necked Phalarope. M. Rubega, D. Schamel, & D. Tracy. 2000. B.O.N.A. 538. 28 pp.
Red Phalarope. D. M. Tracy, D. Schamel, and J. Dale. 2003. B.O.N.A. 698. 32 pp.
Pomarine Jaeger. R. Haven Wiley and David S. Lee. 2000. B.O.N.A. 483. 24 pp.
Parasitic Jaeger. R. Haven Wiley and David S. Lee. 1999. B.O.N.A. 445. 28 pp.
Long-tailed Jaeger. R. Haven Wiley and David S. Lee. 1998. B.O.N.A. 365. 24 pp.
Black-legged Kittiwake. Pat Herron Baird. 1994. B.O.N.A. 92. 28 pp.
Sabine's Gull. R. H. Day, I. J. Stenhouse, & H. G. Gilchrist. 2001. B.O.N.A. 593. 32 pp.
Bonaparte's Gull. J. Burger and M. Gochfeld. 2002. B.O.N.A. 634. 24 pp.
Little Gull. Peter J. Ewins and D. Chip Weseloh. 1999. B.O.N.A. 428. 20 pp.
Laughing Gull. Joanna Burger. 1996. B.O.N.A. 225. 28 pp.
Franklin's Gull. Joanna Burger and Michael Gochfeld. 1994. B.O.N.A. 116. 28 pp.
Mew Gull. W. Moskoff and L. R. Bevier. 2003. B.O.N.A. 687. 28 pp.
Ring-billed Gull. John P. Ryder. 1993. B.O.N.A. 33. 28 pp.
California Gull. David W. Winkler. 1996. B.O.N.A. 259. 28 pp.
Herring Gull. R. J. Pierotti and T. P. Good. 1994. B.O.N.A. 124. 28 pp.
Iceland Gull and Thayer's Gull. R. R. Snell. 2003. B.O.N.A. 699. 36 pp.
Glaucous-winged Gull. N. A. M. Verbeek. 1993. B.O.N.A. 59. 20 pp.
Glaucous Gull. H. Grant Gilchrist. 2001. B.O.N.A. 573. 32 pp.
Great Black-backed Gull. Thomas P. Good. 1998. B.O.N.A. 330. 32 pp.
Least Tern. B. Thompson, J. Jackson, J. Burger, L. Hill, E. Kirsch, & J. Atwood. 1997. B.O.N.A. 290. 32 pp.

Caspian Tern. Francesca J. Cuthbert and Linda R. Wires. 1999. B.O.N.A. 403. 32 pp.
Black Tern. Erica H. Dunn and David J. Agro. 1995. B.O.N.A. 147. 24 pp.
Common Tern. I. C. T. Nisbet. 2002. B.O.N.A. 618. 40 pp.
Forster's Tern. M. K. McNicholl, P. E. Lowther, & J. A. Hall. 2001. B.O.N.A. 595. 24 pp.
Belted Kingfisher. Michael J. Hamas. 1994. B.O.N.A. 84. 16 pp.
Northern Rough-winged Swallow. Michael J. DeJong. 1996. B.O.N.A. 234. 24 pp.
Bank Swallow. Barrett A. Garrison. 1999. B.O.N.A. 414. 28 pp.
Cliff Swallow. Charles R. Brown and Mary B. Brown. 1995. B.O.N.A. 149. 32 pp.
Sedge Wren. J. R. Herkert, D. E. Kroodsma, & J. P. Gibbs. 2001. B.O.N.A. 582. 20 pp.
Marsh Wren. Donald E. Kroodsma & Jared Verner. 1997. B.O.N.A. 308. 32 pp.
American Dipper. Hugh E. Kingery. 1996. B.O.N.A. 229. 28 pp.
Prothonotary Warbler. Lisa J. Petit. 1999. B.O.N.A. 408. 24 pp.
Northern Waterthrush. Stephen W. Eaton. 1995. B.O.N.A. 182. 20 pp.
Louisiana Waterthrush. W. Douglas Robinson. 1995. B.O.N.A. 151. 20 pp.
Common Yellowthroat. Michael J. Guzy and Gary Ritchison. 1999. B.O.N.A. 448. 24 pp.
Le Conte's Sparrow. Peter E. Lowther. 1996. B.O.N.A. 224. 16 pp.
Sharp-tailed Sparrow. Jon S. Greenlaw and James D. Rising. 1994. B.O.N.A. 112. 28 pp.
Swamp Sparrow. Thomas B. Mowbray. 1997. B.O.N.A. 279. 24 pp.
Bobolink. Stephen G. Martin and Thomas A. Gavin. 1995. B.O.N.A. 176. 24 pp.
Red-winged Blackbird. Ken Yasukawa and William A. Searcy. 1995. B.O.N.A. 184. 28 pp.
Yellow-headed Blackbird. Daniel J. Twedt and R. D. Crawford. 1995. B.O.N.A. 192. 28 pp.

Other References

Ailes, I. W. 1980. Breeding biology and habitat use of the upland sandpiper in central Wisconsin. *Passenger Pigeon.* 42:53–63.
Allen, J. N. 1980. The ecology and behavior of the long-billed curlew in southeastern Washington. Am. Ornithol. Union, *Wildlife Monograph* 73:1–64.
Allen, R. P. 1952. *The Whooping Crane.* Research Report No. 2, New York: National Audubon Society.
Bailey, P. F. 1977. The breeding biology of the black tern (*Chlidonias niger surinamensis* Gmelin). M.S. thesis, State University of Wisconsin, Oshkosh.
Bakus, G. W. 1959. Observations on the life history of the dipper in Montana. *Auk* 76:190–207.
Baird, P. A. 1976. Comparative ecology of California and ring-billed gulls (*Larus californicus* and *L. delawarensis*). Ph.D. diss., University of Montana.
Banko, W. 1960. The trumpeter swan. U.S. Fish and Wildlife Service, *North American Fauna* 63:1–214.
Bent, A. C. 1907. The marbled godwit on its breeding grounds. *Auk* 24: 160–67.
Bergman, R. D., P. Swain, and M. W. Weller. 1970. A comparative study of nesting Forster's and black terns. *Wilson Bulletin* 82: 435–44.

Bicak, T. K. 1977. Some eco-ethological aspects of a breeding population of long-billed curlews (*Numenius americanus*) in Nebraska. M.A. thesis. Univ. of Nebraska at Omaha.

Bomberger, M. F. 1982. Aspects of the breeding biology of Wilson's phalarope in western Nebraska. M.S. thesis, Univ. Nebraska, Lincoln. 102 pp.

_____. 1984. Nesting habitat of the Wilson's phalarope in western Nebraska. *Wilson Bull.* 96:126-8.

Boyd, R. L. 1972. Breeding biology of the snowy plover at Cheyenne Bottoms Waterfowl Management Area, Barton County, Kansas. M.S. thesis, Emporia State Univ., Emporia, KS.

Brakhage, G. K. 1965. Biology and behavior of tub-nesting Canada geese. *Journal of Wildlife Management* 29:751–71.

Brown, M. B., J. G. Jorgensen and S. E. Rehme. 2008. Endangered species responses to natural habitat declines Nebraska's interior least terns and piping plovers nesting in a human-created habitat. *Nebraska Bird Review* 76:72–80.

Bunni, M. K. 1959. The killdeer, *Charadrius v. vociferus* Linnaeus, in the breeding season: Ecology, behavior and the development of homiothermism. Ph.D. diss., University of Michigan.

Burger, J. 1974. Breeding adaptations of Franklin's gulls (*Larus pipixcan*) to a marsh habitat. *Animal Behaviour* 22:521–67.

_____, and L. M. Miller. 1977. Colony and nest site selection in white-faced and glossy ibises. *Auk* 94:664–75.

Carter, B. C. 1958. The American goldeneye in central New Brunswick. Canadian Wildlife Service, *Wildlife Management Bulletin,* series 2, no. 9. pp. 1–47.

Chamberlain, M. L. 1977. Observations on the red-necked grebe nesting in Michigan. *Wilson Bulletin* 89:33–46.

Cink, C. 1973. The yellow rail in Nebraska. *Nebraska Bird Review* 41: 24–27.

Cornwell, G. W. 1963. Observations on the breeding biology and behavior of a nesting population of belted kingfishers. *Condor* 65:426–31.

Coulter, M. W., and W. R. Miller. 1968. Nesting biology of black ducks and mallards in northern New England. *Vermont Fish and Game Department Bulletin,* no. 68–2, pp. 1–74.

Crawford, R. D. 1977. Polygynous breeding of short-billed marsh wrens. *Auk* 94:359–62.

Davis, C. A., and L. M. Smith. 1998. Ecology and management of migrant shorebirds in the playa lakes region of Texas. *Wildlife Monographs* No. 140.

_____, and _____. 2001. Foraging strategies and niche dynamics of coexisting shorebirds at stopover sites in the southern Great Plains. *Auk* 118: 484–495.

Dinsmore, S. J. 2012. Birding the Lake McConaughy area. *Prairie Fire* 6(1): 12. 13. 23.

Drewien, R. C. 1973. Ecology of Rocky Mountain greater sandhill cranes. Ph.D. diss., University of Idaho.

Ducey. J. E. 1999. History and status of the trumpeter swan in the Nebraska Sand Hills. *North American Swans* 28:31–39.

DuMont, P. A., and M. H. Swenk. 1934. A systematic analysis of the measurements of 404 Nebraska specimens of geese of the *Branta canadensis* group, formerly contained in the D. H. Talbot collection. *Nebraska Bird Review* 2:1-3–116.

Dunstan, T. C. 1973. The biology of ospreys in Minnesota. *Loon* 45:108–13.

_____, J. E. Mathisen, and J. G. Harper. 1975. The biology of bald eagles in Minnesota. *Loon* 47:5–10.

Eaton, S. W. 1958. A life history study of the Louisiana waterthrush. *Wilson Bulletin* 70:211–236.

Emlen, J. T., Jr. 1954. Territory, nest building and pair formation in the cliff swallow. *Auk* 71:16–35.

Engeling, G. A. 1950. Nesting habits of the mottled duck (*Anas fulvigula maculosa*) in Colorado, Fort Bent and Brazoria counties, Texas. M.S. thesis, Texas A. & M. College.

Erskine, A. J. 1972. *Buffleheads*. Canadian Wildlife Service Monograph Series, no. 4. Ottawa: Information Canada.

Faaborg, J. 1976. Habitat selection and territorial behavior of the small grebes of North Dakota. *Wilson Bulletin* 88:390–99.

Fellows, S. D., and S. I. Jones. 2009. *Status Assessment and Conservation Action Plan for the Long-billed Curlew* (*Numenius americanus*). Washington, D.C.: Dept. of Interior, U.S. Fish & Wildlife Service Bio. Tech, Publ. FWS/BTP-R6012-2009.

Ferguson, R. S., and S. G. Sealey. 1983. Breeding ecology of the horned grebe, *Podiceps auritus,* in southwestern Manitoba. *Canadian Field-Naturalist* 97:4–1–8.

Fitzner, J. N. 1978. The ecology and behavior of the long-billed curlew (*Numenius americanus*) in southeastern Washington. Ph.D. diss., Washington State University.

Fjeldsa, J. 1973. Antagonistic and heterosexual behaviour of the horned grebe, *Podiceps auritus. Sterna* 12:161–217.

Frederickson, L. H. 1970. Breeding biology of American coots in Iowa. *Wilson Bulletin* 82:445–57.

_____. 1971. Common gallinule breeding biology and development. *Auk 88:* 914–19.

Gibson, F. 1971. The breeding biology of the American avocet (*Recurvirostra americana*) in central Oregon. *Condor* 73:444–54.

Gill, K., and P. A. Johnsgard. 2010. The whooping cranes: Survivors against all odds. *Prairie Fire,* Sept., 2010, pp. 12, 13. 16.22. http://www.prairiefirenewspaper.com/2010/9/the-whooping-cranes-survivors-against-all-odds

Girard, G. L. 1939. Notes on life history of the shoveller. *North American Wildlife Conference Transactions* 4:363–71.

_____. 1941. The mallard, its management in western Montana. *Journal of Wildlife Management* 5:233–59.

Glover, F. A. 1953. Nesting ecology of the pied-billed grebe in northwestern Iowa. *Wilson Bulletin* 65:32–39.

Godfrey, R. S. 1975. Behavior and ecology of American woodcock on the breeding range in Minnesota. Ph.D. diss., Univ. of Minnesota, Minneapolis.

Goodwin, R. A. 1960. A study of the ethology of the black tern, *Chlidonias niger surinamensis.* Ph.D. diss., Cornell University.

Green, R. 1976. Breeding behaviour of ospreys *Pandion haliaetus* in Scotland. *Ibis* 118:475–90.

Gregory, C. 2010. Long-billed curlew: Mysterious bird of the Sandhills. *Nebraskaland* 88(4): 32–37.

Grice, D., and J. P. Rogers. 1965. *The Wood Duck in Massachusetts.* Massachusetts Division of Fisheries and Game, Final Report, Project W-19-R.

Gullion, G. W. 1954. The reproductive cycle of American coots in California. *Condor* 71:366–412.

Hahn, H, W. 1950. Nesting behavior of the American dipper in Colorado. *Condor* 52:4962.

Hamilton, R. C. 1975. Comparative behavior of the American avocet and the black-necked stilt (Recurvirostridae). *A. O. U. Monographs* 17: 1–98.

Hammerstrom, F. 1986. *Harrier: Hawk of the Marshes*. Washington, D.C.: Smithsonian Inst. Press.

Hardy, J. 1957. The least tern in the Mississippi Valley. *Michigan State University Museum Publications, Biological Series* I(1): 1–60.

Hays, H. 1973. Polyandry in the spotted sandpiper. *Living Bird* 11:43–57.

Higgins, K. F., and L. M. Kirsch. 1975. Some aspects of the breeding biology of the upland sandpiper in North Dakota. *Wilson Bulletin* 87:96–102.

Hines, J. E. 1977. Nesting and brood ecology of lesser scaup at Waterhen Marsh, Saskatchewan. *Canadian Field-Naturalist* 91:248–55.

Hochbaum, A. H. 1944. *The Canvasback on a Prairie Marsh*. Harrisburg: Stackpole.

Hofslund, P. B. 1959. A life history of the yellowthroat, *Geothlypis trichas*. *Proceedings of the Minnesota Academy of Science* 27:144–74.

Hohn, E. O. 1967. Observations on the breeding biology of Wilson's phalarope (*Steganopus tricolor*) in central Alberta. *Auk* 84:220–44.

Jenni, D. A. 1969. A study of the ecology of four species of herons during the breeding season at Lake Alice, Alachua County, Florida, *Ecological Monographs* 39:245–70.

Johnson, D., and L. D. Igle (compilers). varied dates. Effects of management practices on grassland birds. Northern Prairie Wildlife Research Center, Jamestown, ND. Includes separate accounts for American bittern, marbled godwit, long-billed curlew, willet, Wilson's phalarope, upland sandpiper, northern harrier, sedge wren, Le Conte's sparrow, Nelson's sharp-tailed sparrow and bobolink.

Johnsgard, P. A. 1961. Wintering distribution changes in mallards and black ducks. *American Midland Naturalist*, 66:477–484

_____. 1967. Sympatry changes and hybridization incidence in mallards and black ducks. *American Midland Naturalist*, 77:51–63.

_____. 1980a. Migration schedules of non-passerine birds in Nebraska. *Nebraska Bird Review*, 48:26-36.

_____. 1980b. Migration schedules of passerine birds in Nebraska. *Nebraska Bird Review*, 48:46-57.

_____. 1999a. Where have all the curlews gone? Pp. 139-146. *in* P. Johnsgard, *Earth, Water & Sky: A Naturalist's Stories and Sketches*. Austin: Univ. of Texas Press.

_____. 1999b. The geese from beyond the north wind. Pp. 148-157, *in* P. Johnsgard, *Earth, Water & Sky: A Naturalist's Stories and Sketches*. Austin: Univ. of Texas Press.

_____. 2006. Recent changes in winter bird numbers at Lincoln, Nebraska. *Nebraska Bird Review* 74:16–22.

_____. 2010. Snow geese of the Great Plains. *Prairie Fire*, February, 2010, pp. 12-15. http://www.prairiefirenewspaper.com/2010/3/

_____, and K. Gill. 2011. Sandhill cranes: Nebraska's avian ambassadors at large. *Prairie Fire*, March. 2011, pp. 14, 15, 20. http://www.prairiefirenewspaper.com/2011/02/sandhill-cranes-our-avian-ambassadors-at-large

Jorgensen, J. G., and P. Dunbar, 2005. Multiple black-necked stilt nesting records in the Rainwater Basin. *Nebraska Bird Review* 73: 115-118.

Jorgensen, J. G., S. K. Wilson, J. J. Dinan, S. E. Rehme, S. E. Steckler and M. J. Panella. 2010. A review of modern bald eagle (*Haliaeetus leucocephalus*) nesting records and breeding status in Nebraska. *Nebraska Bird Review* 78:121–126.

Kangarise, C. M. 1979. Breeding biology of Wilson's phalarope in North Dakota. *Bird-banding* 50: 12–22.

Kaufmann, G. W. 1971. Behavior and ecology of the sora, *Porzana carolina,* and Virginia rail. *Rallus limicola.* Ph.D. diss., University of Minnesota, Minneapolis.

Kingsbury, E. W. 1933. The status and natural history of the bobolink *Dolichonyx oryzivorus.* Ph.D. diss., Cornell University. Ithaca, NY.

Kirsch, E. M. 1992. Habitat selection and productivity of least terns (*Sterna antillarum*) on the lower Platte River, Nebraska. Ph.D. diss., Univ. of Montana, Missoula.

Lancaster, D. A. 1970. Breeding behavior of the cattle egret in Colombia. *Living Bird* 9:167–93.

Littlefield, C. D., and R. A. Ryder. 1968. Breeding biology of the greater sandhill crane on Malheur National Wildlife Refuge, Oregon. *Transactions of the North American Wildlife and Natural Resources Conference* 33:444–54.

Low, J. B. 1941. Nesting of the ruddy duck in Iowa. *Auk* 58: 506–17.

_____. 1945. Ecology and management of the redhead, *Nyroca americana,* in Iowa. *Ecological Monographs* 15:35–69.

Ludwig, J. P. 1965. Biology and structure of the Caspian tern (*Hydroprogne caspia*) population of the Great Lakes from 1896–1964. *Bird-Banding* 36: 217–33.

Lunk, W. A. 1962. The rough-winged swallow *Stelgidopteryx ruficollis* (Vieillot); A study based on its breeding biology in Michigan. *Publications of the Nuttall Ornithological Club,* no. 4, pp. 1–155.

McAllister, N. M. 1958. Courtship, hostile behavior, nest establishment, and egg-laying in the eared grebe. *Auk* 75:290–311.

McIntyre, J. W. 1988. *The Common Loon: "Spirit of Northern Lakes."* Minneapolis: Univ. of Minnesota Press.

McKinney, F. 1965. The displays of the American green-winged teal. *Wilson Bulletin* 77:112–21.

McNicholl, M. K. 1971. The breeding biology and ecology of Forster's tern (*Sterna forsteri*) at Delta, Manitoba, M.S. thesis, University of Manitoba.

Martin, S. G. 1971. Adaptations for polygynous breeding in the bobolink. *American Zoologist* 14:109–19.

Meanley, B. 1955. A nesting study of the little blue heron in eastern Arkansas. *Wilson Bulletin* 67:84–99.

_____. 1963. Prenesting activity of the purple gallinule near Savannah, Georgia, *Auk* 80:545–47.

_____. 1969. *Natural History of the King Rail.* U.S. Fish and Wildlife Service, North American Fauna, no. 67. Washington, D.C.

_____, and A. G. Meanley. 1959. Observations on the fulvous whistling duck in Louisiana *Wilson Bulletin* 71:31–45.

Mendall, H. L. 1958. The ring-necked duck in the Northeast. *University of Maine Bulletin,* vol. 60, no. 16; and *University of Maine Studies.* 2nd ser., no. 73: 1–317.

Mitchell, R. M. 1977. Breeding biology of the double-crested cormorant at Utah Lake. *Great Basin Naturalist* 37: 1–23.

Mock, D. M. 1976. Pair-formation displays of the great blue heron. *Wilson Bulletin* 88: 185–230.

Morse, T. E., J. L. Jakabosky, and V. P. McCrow. 1969. Some aspects of the breeding biology of the hooded merganser. *Journal of Wildlife Management* 33:596–604.

Mousley, H. 1939. Home life of the American bittern. *Wilson Bulletin* 51:83–85.

Murray, B. G., Jr. 1969. A comparative study of the Le Conte's and sharp-tailed sparrows. *Auk* 86:199–231.

_____. 1983. Notes on the breeding biology of Wilson's phalarope. *Wilson Bull.* 95:472–475.

Nero, R. E. 1956. A behavior study of the red-winged blackbird. *Wilson Bulletin* 68:5–37, 129–50.

Noble, G. K., M. Wurm, and M. Schmidt. 1938. Social behavior of the black-crowned night heron. *Auk* 55:7–40.

Nowicki, T. 1973. A behavioral study of the marbled godwit in North Dakota. M.S. thesis, Central Michigan University.

Nuechterlein, G. 1975. Nesting ecology of western grebes on the Delta Marsh, Manitoba. M.S. thesis, Colorado State University.

_____. 1981a. Courtship and reproductive isolation between western grebe morphs. *Auk* 98:335–49.

_____. 1981b. Variation and multiple functions of the advertising display of western grebes. *Behaviour* 76:289–318.

Olson, S. T., and W. H. Marshall. 1952. The common loon in Minnesota. *Occasional Papers of the Minnesota Museum of Natural History,* no. 5, pp. 1–77.

Orians, G. H., and G. M. Christman. 1968. A comparative study of the behavior of red-winged, tricolored, and yellow-headed blackbirds. *University of California Publications in Zoology* 84: 1–85.

Oring, L. W. 1969. Summer biology of the gadwall at Delta, Manitoba. *Wilson Bulletin* 81:44–54.

_____, and M. L. Knudson. 1973. Monogamy and polyandry in the spotted sandpiper. *Living Bird* 2:59–73.

Palmer, R. S. 1941. A behavior study of the common tern. *Proceedings of the Boston Society of Natural History* 42:1–119.

Peek, F. W. 1971. Seasonal change in the breeding behavior of the male red-winged blackbird. *Wilson Bulletin* 83:383–95.

Peterson, A. J. 1955. The breeding cycle in the bank swallow. *Wilson Bulletin* 67:235–86.

Peyton, M. M. 2012. The Kinsley eagles. *Prairie Fire* 6(1): 1.3.

Phillips, R. S. 1972. Sexual and agonistic behavior in the killdeer (*Charadrius vociferus*). *Animal Behavior* 20:1–9.

Poole, A. F. 1989. *Ospreys: A Natural and Unnatural History.* Cambridge, UK: Cambridge Univ. Press.

Pospichal, L. 8., and Marshall, W. H. 1954. A field study of the sora rail and Virginia rail in central Minnesota. *Flicker* 26:2–32.

Poston, H. J. 1969. Home range and breeding biology of the shoveler. M.S. thesis, Utah State University, Logan, Utah.

Pratt, H. M. 1970. Breeding biology of great blue herons and common egrets in central California. *Condor* 72:407–16.

Purdue, J. R. 1976. Adaptation of the snowy plover on the Great Salt Plains, Oklahoma. *Southwestern Naturalist* 21: 347–57.

Ratti, J. T. 1979. Reproductive separation and isolating mechanisms between dark- and light-phase western grebes. *Auk* 96: 573–86.

Rawls, C. K., Jr. 1949. An investigation of life history of the white-winged scoter (*Melanitta fusca deglandi*). M.S. thesis, University of Minnesota.

Redmond, R. L, and D. A. Jenni. 1986. Population ecology of the long-billed curlew (*Numenius americanus*) in western Idaho. *Auk* 103: 755–67.

Rogers, J. A., Jr. 1977. Breeding displays of the Louisiana heron. *Wilson Bulletin* 89: 266–85.

Ryan, M. R., R. B. Renken, and J. J. Dinsmore. 1984. Marbled godwit habitat selection in the northern prairie region. *J. Wildl. Mgmt.* 48: 1206–18.

Samuel, D. E. 1971. The breeding biology of barn and cliff swallows in West Virginia. *Wilson Bulletin* 83:284–30.

Sauer, E. G. F 1963. Migration habits of golden plovers *Proc. XIII Internat. Ornithol. Congr.* 454–467.

Schaller, G. B. 1964. Breeding behavior of the white pelican at Yellowstone Lake, Wyoming. *Condor* 66:3–23.

Sheldon, W. G. 1967. *The Book of the American Woodcock.* Amherst: University of Massachusetts Press.

Sherrod, S. K., C. M. White, and F. S. L. Williamson. 1976. Biology of the bald eagle on Amchitka Island, Alaska. *Living Bird* 15:143–82.

Sidle, J. G., *et al.* 1991. The 1991 survey of least terns and piping plovers in Nebraska. *Nebraska Bird Review* 59:133–149.

Silcock, W. R. 2006. White-cheeked geese in Nebraska. *Nebraska Bird Review* 74:99–104.

Simmons, R. E. 1983. Polygyny, ecology and mate choice in the northern harrier, *Circus cyaneus* (L.). M.S. thesis, Acadia Univ., Wolfville, NS. 177 pp.

Skagen, S. K., and F. L. Knopf. 1994a. Migrating shorebirds and habitat dynamics at a prairie wetland complex. *Wilson Bull.* 106: 91–105.

_____, and _____. 1994b. Residency patterns of migrating shorebirds at a midcontinental stopover. *Condor* 96: 949–958.

Spencer, H. E., Jr. 1953. The cinnamon teal (*Anas cyanoptera* Vieillot): Its life history, ecology, and management. M.S. thesis, Utah State University.

Stewart, R. E. 1953. A life history study of the yellow-throat. *Wilson Bulletin* 65:99–115.

Storer, R. W. 1969. Behavior of the horned grebe in spring. *Condor* 71:180–205.

_____, and G. L. Nuechterlein. 1985. An analysis of plumage and morphological characters of the two color forms of the western grebe. *Auk* 102: 109–119.

Tanner, W. D., Jr., and Hendrickson, G. O. 1954. Ecology of the Virginia rail in Clay County, Iowa. *Iowa Bird Life* 24:65–70.

_____. 1956. Ecology of the king rail in Clay County, Iowa. *Iowa Bird Life* 26:54–56.

Tate, D. J. 1973. Habitat usage by the chipping sparrow (*Spizella passerina*) in northern Lower Michigan. Ph.D. diss., University of Nebraska.

Terrill, L. 1943. Nesting habits of the yellow rail in Gaspe County, Quebec. *Auk* 60: 171–180.

Tinbergen, N. 1959. Comparative studies of the behaviour of gulls (Laridae): A progress report. *Behaviour* 15: 1–70.

Todd, R. L. 1977. Black rail. In *Management of Migratory Shore and Upland Game Birds in North America,* pp. 71–83. Washington D. C.: International Association of Fish and Wildlife Agencies.

Tomlinson, D. N. S. 1976. Breeding behaviour of the great white egret. *Ostrich* 47:161–78.

Tompkins, I. R. 1959. Life history notes on the least tern. *Wilson Bulletin* 71: 313–22.

———. 1965. The willets of Georgia and South Carolina. *Wilson Bulletin* 77: 151–67.

Trauger, D. L. 1971. Population ecology of lesser scaup (*Aythya affinis*) in subarctic taiga. Ph.D. diss., Iowa State University.

Trautman, M. B., and S. I. Clines. 1964. A nesting of the purple gallinule in Ohio. *Auk* 81:224–26.

Tuck, L. M. 1972. *The Snipes: A Study of the genus* Capella. Canadian Wildlife Service, Monograph Series no. 5. Ottawa.

Vermeer, K. 1970. Breeding biology of California and ring-billed gulls: A study of ecological adaptation to the inland habitat. Canadian Wildlife Service, Report Series no. 12. Ottawa.

Verner, 1. 1965. Breeding biology of the long-billed marsh wren. *Condor* 67:6–30.

Walkinshaw, L. H. 1935. Studies of the short-billed marsh wren in Michigan. *Auk* 52: 362–69.

———. 1953. Life history of the prothonotary warbler. *Wilson Bulletin* 65:152–68.

Watson, A. T. 1977. *The Hen Harrier*. Berkhamstead, England: T. & A. D. Poyser.

Weller, M. W. 1961. Breeding biology of the least bittern. *Wilson Bulletin* 73:11–35.

White, H. C. 1953. The eastern belted kingfisher in the Maritime Provinces. *Fisheries Research Board of Canada Bulletin* 97: 1–44.

———. 1957. Food and natural history of mergansers on salmon waters in the Maritime Provinces of Canada. *Fisheries Research Board of Canada Bulletin* 116: 1–63.

Wiese, J. H. 1976. Courtship and pair formation in the great egret. *Auk* 93:709–24.

Wilcox, L. R. 1959. A twenty year banding study of the piping plover. *Auk* 76: 129–52.

Willson, M. F. 1964. Breeding ecology of the yellow-headed blackbird. *Ecological Monographs* 36:51–77.

Woolfenden, G. E. 1956. Comparative breeding behavior of *Ammospiza caudacuta* and *A. maritima*. *Univ. of Kansas Museum Publications* 10: 45–75.

Zimmerman, J. L. 1966. Polygyny in the dickcissel. *Auk* 83:534–46.

The University of Nebraska–Lincoln does not discriminate
based on gender, age, disability, race, color,
religion, marital status, veteran's status,
national or ethnic origin,
or sexual orientation.

www.ingramcontent.com/pod-product-compliance
Lightning Source LLC
Chambersburg PA
CBHW020606270326
41927CB00005B/197